THE CASE FOR
PLUTO

Al Tombaugh (foreground) and his wife, Cherylee Tombaugh (to his right), were among the Pluto defenders who gathered at New Mexico State University a week after the planet's reclassification by the International Astronomical Union. Al Tombaugh is the son of the late Clyde Tombaugh, who discovered Pluto in 1930.

THE CASE FOR
PLUTO

How a Little Planet Made a Big Difference

ALAN BOYLE

WILEY

John Wiley & Sons, Inc.

Copyright © 2010 by Alan Boyle. All rights reserved

Published by John Wiley & Sons, Inc., Hoboken, New Jersey
Published simultaneously in Canada

Design by Forty-five Degree Design LLC

For general information about our other products and services, please contact our Customer Care Department within the United States at (800) 762-2974, outside the United States at (317) 572-3993 or fax (317) 572-4002.

Wiley also publishes its books in a variety of electronic formats. Some content that appears in print may not be available in electronic books. For more information about Wiley products, visit our web site at www.wiley.com.

Library of Congress Cataloging-in-Publication Data:
Boyle, Alan, date.
    The case for Pluto: how a little planet made a big difference/Alan Boyle.
        p. cm.
  Includes bibliographical references and index.
  ISBN 978-0-470-50544-1 (cloth)
1.   Pluto (Dwarf planet)   2.   Solar system.   I.   Title.
  QB701.B69 2009
  523.49'22—dc22

                                                   2009015961

Printed in the United States of America

10 9 8 7 6 5 4 3 2 1

To Tonia, my star

All that I know
Of a certain star
Is, it can throw
(Like the angled spar)
Now a dart of red,
Now a dart of blue;
Till my friends have said
They would fain see, too,
My star that dartles the red and the blue!
Then it stops like a bird; like a flower hangs furled:
They must solace themselves with the Saturn above it.
What matter to me if their star is a world?
Mine has opened its soul to me, therefore I love it.

—*Robert Browning, "My Star"*

CONTENTS

ACKNOWLEDGMENTS

Like many other observers, I assumed that the International Astronomical Union (IAU) would have the final word on what is and isn't a planet. And I said as much in my long-running blog on msnbc.com. It took Alan Stern to convince me that I was wrong.

When I wrote in September 2006 that Pluto was gone for good from the nine-planet club, no matter what any dissenter might say, the head of NASA's New Horizons mission to Pluto was quick to send me a stern e-mail in response. "Who is pulling the wool over your eyes?" Stern wrote. I owe Alan Stern a debt of gratitude—first for taking the wool from my eyes, then for sharing his insights on the planetary frontier, and finally for reviewing the manuscript.

Mark Sykes, director of the Planetary Science Institute, was even more thorough in his review, as befits an accomplished scientist who happens to be an attorney as well. My third reviewer was Annette Tombaugh-Sitze, the daughter of Pluto's discoverer, who was kind enough to share family pictures as well as family reminiscences.

I'm grateful to all three for their encouragement, corrections, and comments. Any errors that remain in this book are mine alone.

In addition to these three advisers, this book draws upon interviews with a dream team of astronomers and planetary scientists. I thank these sources for sharing their expertise:

- Brian Marsden of the Harvard-Smithsonian Center for Astrophysics, director emeritus of the IAU's Minor Planet Center;
- Owen Gingerich, professor emeritus of astronomy and the history of science at Harvard University, as well as a senior astronomer emeritus at the Smithsonian Astrophysical Observatory;
- Steven Soter, scientist-in-residence at New York University and research associate at the American Museum of Natural History;
- Edward Bowell, astronomer at Lowell Observatory and president of IAU Division III for planetary systems sciences;
- Dana Backman, astronomer at the SETI Institute and coauthor of the textbook *Perspectives on Astronomy*;
- Astronomers Thomas T. Arny and Stephen Schneider, authors of *Pathways to Astronomy*;
- German journalist Daniel Fischer, creator of the online journal *The Cosmic Mirror* and a witness to the planetary politics in Prague.

This book also draws upon interviews conducted over the past three years in the course of my work as science editor for msnbc.com. I'm particularly grateful for the insights provided by Mike Brown, planetary astronomer at the California Institute of Technology (and dwarf-planet discoverer extraordinaire); Alan Boss, theoretical astrophysicist at the Carnegie Institution of Washington; Neil deGrasse Tyson, director of the Hayden Planetarium; and David Weintraub, astronomer at Vanderbilt University.

I want to thank my old friend Doug "Chief" Slauson for taking me on a tour of the planets in Mount Vernon, Iowa, and showing me his pride and joy, the Eastern Iowa Observatory and Learning Center.

Thanks also to my editor at msnbc.com, Michael Wann, for his interest and patience as I slogged through this extracurricular project. I also appreciate the support of Jennifer Sizemore, vice president and editor in chief of msnbc.com.

Several people helped round up the images accompanying the text, including Antoinette Beiser and Steele Wotkyns at Lowell Observatory, Ray Villard at the Space Telescope Science Institute, Lars Lindberg Christensen at the International Astronomical Union, Dan Durda and Eliot Young at the Southwest Research Institute, Geoff Chester at the U.S. Naval Observatory, Darren Phillips at New Mexico State University, and Carolina Martinez at NASA's Jet Propulsion Laboratory.

I am immensely grateful to Stephen Power, my editor at John Wiley & Sons, who contacted me about taking on *The Case for Pluto*. This book exists primarily because he gave me the opportunity to write it. I also thank Mel Berger, my literary agent at William Morris Agency, who helped me take advantage of that opportunity. I could ask for no better team to guide me through the writing of my first honest-to-goodness book. Thanks also to production editor Rachel Meyers, freelance copy editor Roland Ottewell, and the rest of the sharp-eyed team at Wiley.

Finally, I'm thankful for my family's love and support, especially from my children, Natalie and Evan, and my wife, Tonia, to whom this book is dedicated.

1

THE PLANET IN THE CORNFIELD

Out amid the cornfields of Iowa, my friend Chief built a monument to Pluto, the picked-on planet.

The red, oval-shaped plaque was smaller than a stop sign, with a pimple of polished steel sticking up from the surface to represent Pluto's size. It was mounted on a metal pole by the side of a blacktop road, four miles west of the town of Mount Vernon, population 3,628.

Chief, who got his nickname during childhood because he was part Native American, was one of my best friends in high school. Now he works at the University of Iowa and has become an amateur astronomer of some repute. A few years ago, he and other volunteers started up a group called the Mount Vernon Solar Tourist Society and erected the plaques just for fun, to show how big and empty our solar system is.

You can't understand the distances that separate the planets just by looking at a schoolroom poster. They're usually displayed right next to each other like some kind of celestial police lineup, with pea-sized Pluto pictured right alongside his big brothers Uranus and Neptune.

To provide a better sense of scale, folks like Chief have laid out scores of mini–solar systems around the world. It's the best way to relate the size of the planets to the immense distances involved. For example, the scale model in Washington, D.C., has a five-inch-wide sun, and plaques depicting the planets are lined up along the National Mall for a third of a mile. Boston's planetary parade extends more than 9 miles, leading out from an eleven-foot-wide sun. And you'd have to drive more than 180 miles to get from Stockholm's solar stand-in (actually the round-domed Stockholm Globe Arena) to a five-inch-wide Pluto perched on a monument in Sweden's Dellen Lake district.

The Mount Vernon Solar Tourist Society set up a five-foot-wide sun at the city park, and planted plaques leading west along First Avenue, plus an "Asteroid Crossing" road sign next to Cornell College's campus to mark an imaginary asteroid belt. That sign shows up in a fair number of pictures on the

Internet, but few if any of the photographers have figured out that it's actually part of a set.

Each of the nine planetary plaques displays a list of facts about the world in question, and includes a scaled-down circle of bright steel to represent the planet's relative size. Pluto's circle was about as small as the artist could make it, as big around as a pebble that gets caught between your tires on a gravel road.

Even when the plaques were being put up, the society was having second thoughts about Pluto. One of the inscriptions on its plaque read, "If Pluto was discovered today, it would not be called a planet, but a minor planet."

Since then, Pluto has suffered putdowns galore. It was left out when New York's American Museum of Natural History remodeled its planetary exhibits. More and more worlds like Pluto were found on the solar system's rim, and in 2005 astronomers determined that one of them was actually bigger than Pluto.

If that newfound world—known at first as Xena (the Warrior Princess) and later named Eris (the Goddess of Discord)—had been accepted into the planetary clan, Chief might have had to add one more monument to the set. It would have been about seven miles out of town, by my calculation. And for a while, it looked as if things were heading in that direction.

A committee charged with settling the question drew up a proposal that would have boosted the solar system's official planet count to twelve, including Eris (née Xena), as well as Pluto's largest moon, Charon, and the asteroid Ceres.

But in 2006 when the proposal came up for a vote by the International Astronomical Union (IAU), the world body that deals with astronomical names and definitions, it was hooted down. Instead, a few hundred astronomers voted to throw Pluto and the other lesser worlds into a different class of celestial objects, known as "dwarf planets."

That might not sound so bad. After all, a dwarf planet is still a planet, just as a dwarf galaxy is still a galaxy, and just as a dwarf star (like our sun) is still a star. Right?

Wrong.

When the IAU reclassified Pluto, it declared that dwarf planets weren't actually real planets, but mere also-rans in the celestial scheme. That's when things turned ugly.

Astronomers split into opposing camps. One said he had "nothing but ridicule" for the IAU's decision.[1] Another said anyone who disagreed with the decision should be "stomped."[2] Web sites were remade. Textbooks were rewritten. Pluto fans of all ages were heartbroken.

"My ten-year-old daughter is *furious* about this," one parent wrote me.

Seventy-six-year-old Dorothy Timmerman was also hopping mad. "Pluto is my very own personal planet! It was discovered the year I was born!" she wrote. "They can't take my planet away! I want my planet back!"[3]

It was obvious that little Pluto's fate had sparked a huge battle—a battle that ranged far beyond conference halls and observatories, all the way to First Street in Mount Vernon, Iowa.

·   ·   ·

Chief brought me to the city park one night to have a look at the mini-sun, sitting upright on a stone pedestal like a red-orange snow saucer. Nine badges were mounted on the disk, naming each of the planets and their distances from the sun monument.

There was Mercury, Venus, Earth, Mars, Jupiter, Saturn, Uranus, Neptune, and . . . uh-oh. Pluto's badge was still there, but it was defaced with a black "X," drawn with a permanent marker over the lettering.

"Oh dear," Chief remarked. "Someone is too scientifically read up, I'm afraid."

We hopped into Chief's car and drove through Mount Vernon's solar system, checking each plaque. Mercury was a speck of polished metal, shining on a red plaque erected on the next block. Venus and Earth were two more shiny specks on plaques screwed onto brick buildings downtown, a couple of blocks farther west. Mars's speck was on a sign planted right outside the fire station.

We drove past the Asteroid Crossing sign and went half a mile more, to the edge of town. There, mounted on another oval plaque, was the shiny softball-sized disk of metal that stood for Jupiter. Saturn, a disk just slightly smaller, was displayed on a sign next to the old country schoolhouse, a mile out of town on Old Lincoln Highway. Baseball-sized Uranus was more than a mile farther, Neptune another mile and a quarter.

Then we slowed down, watching closely as the car's headlights illuminated the grassy roadside. We went a mile farther. Two miles. Three miles. Nothing.

We turned around and drove back toward town. "Maybe it's just lost in the weeds," Chief suggested hopefully. "I'll go back and look tomorrow, when it's light out."

If Mount Vernon's Pluto was X'd out, Chief didn't think there'd be much sentiment for replacing it. "We can always say, 'Well, there are only eight planets in the solar system,' " he said.

I didn't take much consolation from that. Chief dropped me off at my car and headed home, but I drove down the blacktop again. I carefully counted off the tenths of a mile on the odometer, knowing full well that Pluto should be 1.07 miles beyond Neptune (or about a mile beyond the dead opossum in the middle of the road).

Sure enough, there it was: a flicker of red in the headlights, just beyond an intersection with a gravel road. I pulled off onto the gravel, shone the lights full on the plaque, and ran my finger over that pimple of steel.

Pluto's monument was still the same, out amid the cornfields of Iowa. And Pluto is still the same, out in the celestial countryside—no matter what we say about it on our own somewhat bigger pimple of a planet.

What is it about Pluto that stirs up so much trouble, for its defenders as well as its detractors? What leads Pluto's elementary school fans to write tear-stained letters to astronomers, protesting the snub? Why are some scientists so anxious to remake the solar system, while others leap to the barricades in Pluto's defense?

In part it's because Pluto has always been the oddball of the solar system family. Even before the latest flap, astronomers knew it was totally unlike the other eight planets in the traditional lineup.

First, there's the question of size. When it comes to mass, Earth has 465 times as much as Pluto (and Jupiter has 318 times as much as Earth, by the way). When it comes to diameter, Pluto is smaller than some of the solar system's moons (as is Mercury, by the way).

Pluto's rotational axis is tipped so steeply that for part of its year, the sun rises in the (celestial) south and sets in the north. Its orbit is also tipped—17 degrees from the solar system's ecliptic plane, which would translate to more than a mile in altitude if you extended that angle for four and a half miles outside Mount Vernon. The orbit is so eccentric that Pluto comes closer to the sun than Neptune for twenty years at a time, and then careens out as much as 1.8 billion miles farther away.

The icy world's history is as eccentric as its orbit. Pluto was discovered by a high school graduate, who was following up on claims that turned out to be based on completely wrong assumptions, which were made by an astronomer who was also convinced he saw canals built on Mars. Once the faraway speck's existence was confirmed, the darn thing was named by an eleven-year-old girl, and the planet in turn lent its name to a Disney cartoon dog.

More recently, astronomers learned that Pluto has one moon that could well be considered a planet in its own right (a planetoon?), plus two more mini-moons. They also learned that Pluto isn't your typical snowball: It appears to be covered with frozen nitrogen, carbon monoxide, methane, and ethane—a complex coating that gives rise to a thin atmosphere during Plutonian summer. That atmosphere may settle back down to the surface as frost during the long winter—or it may not. Astronomers are waiting to find out, because they haven't yet had a chance to study Pluto in winter.

All this alone is enough to make Pluto the embarrassing weird uncle of the solar system, wearing a leisure suit with a squirting flower in the lapel. But the most dramatic fall in Pluto's fortunes has come about in the past few years— ironically, because the fortunes of planet hunters have risen so dramatically.

Telescopes have come a long way since 1930, when twenty-four-year-old Clyde Tombaugh discovered Pluto by poring over photographic plates at Arizona's Lowell Observatory. Beginning in the early 1990s, astronomers have picked up the glitter of other specks on the solar system's edge. It soon became widely accepted that, over the course of billions of years, the planet formation process had laid down a ring of icy bits beyond the orbit of Neptune.

Pluto came to be regarded as not only the smallest planet, but also the largest of those icy bits on the edge. And many astronomers said it would be only a matter of time before they found objects way out there that were bigger than Pluto.

That prediction came true in 2005, when Caltech astronomer Mike Brown reported the discovery of Eris, which has an orbit that's even farther away, more eccentric, and more tipped than Pluto's. There's no theoretical reason why other objects bigger than Pluto, or even bigger than Earth, couldn't be lurking out amid the solar system's nether regions.

Pluto turned out to be not quite as special as astronomers originally thought. And a lot of those astronomers thought an object had to be very, very special in order to be called a planet. It wouldn't do to have a list of tens, or even hundreds, of planets to remember.

That's why the IAU pushed through a resolution that required something to be a real standout in order to be dubbed a real planet: It had to "clear the neighborhood of its orbit"—that is, it had to be the biggest thing by far in its orbital space.

Some astronomers supposed that the matter was settled merely because an international body had passed a resolution. "Pluto is dead," Mike Brown told reporters just after the IAU decision. "There are finally, officially, eight planets in the solar system."[4]

Such suppositions turned out to be wrong on several counts. Far from resolving the issue on scientific grounds, the decision sparked a whole series of arguments; first of all over the definition (should specialness be the determining factor?) and the semantics (isn't a dwarf planet still a planet?).

Other arguments suggest that it's still way too early to pigeonhole planets, dwarf or otherwise. One space probe is heading for Pluto, while another is on its way to a dwarf planet closer to home, known as Ceres. Both are due to arrive at their destinations in the year 2015, and they're both expected to reveal that those little worlds are far more complex than scientists think.

At the same time, more worlds are being discovered every month—not just in our own solar system, but also in star systems light-years away. Astronomers have found "hot Jupiters" that are bigger than our own Jupiter, but circle their own suns in orbits that are tighter than that of Mercury, the closest-in of our solar system's planets. They are watching an infant planet still shrouded in gas and dust.[5] They've even spotted alien asteroid belts and ice rings.[6]

In the midst of this revolution in our understanding of how stars and planets are built—a revolution that is revealing a greater diversity of celestial wonders—does it really make sense to lay down a definition of planethood that excludes more than it includes? It might make better sense to widen our perspective and keep an eye out for the seemingly insignificant worlds that just might end up telling us more about the origins of the universe—and perhaps the origins of life as well.

The battle lines in the case for Pluto go far beyond planetary science, to take in some of society's sensitive topics: Are scientific questions decided by a single vote or a resolution, or does it take years of claims and challenges for the answer to emerge? How long do you have to wait? How much weight do you give to the pull of history and culture? Will scientific dogma turn out to be just another type of belief system, where authorities dictate the terminology and the truth? These bigger questions apply not just to Pluto and planethood, but also to issues closer at hand, such as climate change and the skirmishes between religious believers and scientific skeptics.

Then there's Pluto's emotional pull. Some people see the situation as a classic underdog-versus-establishment struggle, or a fight to defend the "first American planet" from foreigners. For kids, Pluto ranks right up there with the little engine that could. "Children love that little planet," Patsy Tombaugh, the widow of Pluto's discoverer, once said.[7]

For others, Pluto's weird plight became the punch line for a string of jokes about life's disappointments—and a reminder that anyone or anything that doesn't live up to expectations can be struck off the A-list, just like that!

"If Pluto can be downgraded, why not demote Duke football to 'dwarf team'?" sports columnist Frank Deford asked.[8]

"An international group of scientists who demoted the planet Pluto to dwarf status . . . met in Oslo, Norway, today and reclassified the Bush White House as a dwarf presidency," humorist Andy Borowitz joked.[9]

We're not strictly talking about science here. But Pluto's story is about more than just science. It's also about the personalities and politics, the parodies and pop culture. You can't leave out those parts of the story if you're going to make the case for Pluto.

When you get right down to it, the case for Pluto doesn't have all that much to do with the fate of Pluto itself. That pimple of a planet won't be affected by any resolution or petition issued back here on Earth. But the debate *does* have everything to do with how we see the universe around us—even if your vantage point is out amid the cornfields of Iowa.

2

FELLOW
WANDERERS

The cosmos seemed so much simpler in ancient times.

The word "planet" traces its origins to the Greek word for "wanderer," but the underlying concept goes back much further. When the earliest humans looked up at the lights in the sky, it didn't take them long to notice that most of them formed unchanging patterns like the Big Dipper and the three-star belt of Orion. Other lights, however, changed their position relative to these fixed stars.

The sun and the moon were the most obvious of these wanderers. Another five points of light—which we now call Mercury, Venus, Mars, Jupiter, and Saturn—moved back and forth through the constellations. The Greeks added these five "wandering stars" to the sun and the moon, making a total of seven *planetoi* circling Earth.[1]

That worldview held sway throughout the Roman era and well into the Middle Ages. But the rise of the Renaissance stirred new curiosities, new questions, new tools, and clever new ways to use those tools to provide answers. In the process, the cosmos became more complex.

In the 1500s and early 1600s, Nicolaus Copernicus and Johannes Kepler laid down theories that turned the old worldview around: Instead of having the sun, moon, and planets circling Earth, they laid out a system in which the planets circled the sun. Earth became just another planet, orbited by the moon. It was a dramatic change in perspective that took us humans and our home out of the universe's central position.

The groundwork for our current basic understanding of the solar system was thus put into place well before the milestone year of 1609, when Galileo Galilei, armed only with the spyglass of his own construction, began making his revolutionary observations of the moon and planets.

You could argue that Galileo was the first man in recorded history to claim the discovery of a new planet. Four of them, in fact. When he looked through his telescope at Jupiter, he saw four specks of light that lined up with the planet's disk and seemed to change position from night to night. Showing the political savvy embodied a century earlier by his fellow Florentine, Niccolò Machiavelli, Galileo named his four finds the "Medicean stars" in honor of his most important patron, Grand Duke Cosimo Il de' Medici.

Galileo gushed that the newfound objects never strayed far from Jupiter, the dominant planet in the prince's horoscope. For that reason, the astronomer told Cosimo, "it appears that the Maker of the Stars himself, by clear arguments, admonished me to call these new planets by the illustrious name of Your Highness above all others."[2]

Ironically, the name didn't stick. Instead, "these new planets" are now known as Jupiter's Galilean satellites, the four biggest moons of our solar system's biggest planet. Using a garden-variety pair of binoculars, you can see them much as Galileo did, particularly from wide-open spaces like the cornfields of Iowa.

Galileo's reports sparked the seventeenth century's grand clash between the medieval view of the heavens, which put Earth in a special place at the center of God's universe, and the revolutionary view that Earth was merely a planet that circled the sun along with other wanderers.

The effect on Galileo's career is well-known: He faced not just one, but two church inquisitions that left him under suspicion of heresy—and under house arrest for the last nine years of his life.[3]

The Medicean stars had a more salutary influence on other scientists who were already working out the implications of the Copernican worldview. Kepler, for instance, saw Galileo's discovery as confirmation that the planets were on a par with Earth, and he was among the first to recognize a distinction between planets that circle the sun and moons that circle the planets. "These four little moons exist for Jupiter, not for us," Kepler wrote. "Each planet in turn, together with its occupants, is served by its own satellites."[4]

The line between planets and their satellites became clearer as time went on. In 1655, Dutch astronomer Christiaan Huygens spotted Saturn's largest moon, Titan. Over the three decades that followed, Italy's Giovanni Domenico Cassini found four more Saturnian moons. By the time the eighteenth century dawned, old-fashioned observers might have reckoned the planet count at sixteen, but when most astronomers took stock of the sky they counted six planets and ten moons.

Even today, the definition for a planet's moon is more clear-cut than the definition for the planet itself. Any object in orbit around the planet would be considered its satellite, and if the object is of natural origin, you'd call it a moon.

The planet debate stirred again in 1781, when William Herschel, a musician turned amateur astronomer, spotted what looked like a comet through the seven-foot-long, home-built telescope he set up in the garden behind his house in the English spa city of Bath. After repeated observations, astronomers across Europe competed to calculate the object's orbit—and the Royal Society's president, Joseph Banks, urged

Herschel to make up his mind about what he had initially guessed was "a comet of a new species, very like a fixed star." Wrote Banks, "Some of our astronomers here incline to the opinion that it is a planet and not a comet; if you are of the opinion it should forthwith be provided with a name or our nimble neighbors, the French, will certainly save us the trouble of baptizing it."[5]

Herschel was persuaded to go with the planetary designation—which was the right choice. Further observations confirmed that his comet was indeed a new planet.

Thanks to his lucky discovery, Herschel became the toast of the scientific world. Britain's King George III, who was grateful for the distraction from his troubles with the American colonies, appointed him his private astronomer. Following Galileo's lead, Herschel named the newfound world Georgium Sidus, or the Georgian Star.

Like Galileo, Herschel laid it on thick. "As a subject of the best of Kings, who is the liberal protector of every art and science; as a native of the country from whence this Illustrious Family was called to the British throne; and as a person now more immediately under the protection of this excellent Monarch and owing everything to his unlimited bounty; I cannot but wish to take this opportunity of expressing my sense of gratitude, by giving the name Georgium Sidus to a star which (with respect to us) first began to shine under His auspicious reign," he wrote.[6]

Once again, the fawning name didn't stick. Instead, Prussian astronomer Johann Bode came up with Uranus, a

Greek name that refers to the mythological father of the god Saturn. That name was more in conformity with the classical tradition for planet names—and besides, it carried less political baggage for international use.

Herschel steadfastly refused to refer to the planet as Uranus, out of loyalty to his king and the kingly name he gave his discovery. Eventually, however, even British astronomers came around, giving generations of schoolchildren a pronunciation to giggle over.

William Herschel was also a central figure in the nineteenth century's biggest planet debate—an eerie foreshadowing of the twenty-first-century debate over Pluto and its ilk.

Back in 1766, German astronomer Johann Titius saw a mathematical pattern in the spacing of the six planetary orbits known at that time. Based on that pattern, he and his colleague Johann Bode (the man who ended up naming Uranus) figured that one additional planet should theoretically fill the gap between Mars and Jupiter. They also said yet another planet should be found beyond Saturn's orbit.

The placement of Uranus fit the pattern, seemingly confirming what was known as the Titius-Bode law. That revved up the search for a previously undetected world between Mars and Jupiter. Sure enough, an Italian monk and astronomer, Giuseppe Piazzi, found a prospect in the predicted orbit in 1801. "The first of January I discovered a star, which by its motion strongly appears to be a planet. . . . I would very much like for you to search for it," Piazzi wrote in a letter to Herschel.[7] Once the find was confirmed, Piazzi named the planet Ceres Ferdinandea, in honor of Ceres, the Roman goddess of the harvest, as well as the Sicilian king Ferdinand IV. Other astronomers quickly shortened the name.

The discovery of Ceres was a classic example showing how even an unfounded theory can sometimes lead to substantive discoveries. To this day, there is no scientific explanation behind the Titius-Bode law; nevertheless, it pointed nineteenth-century astronomers to a previously unknown celestial body. A whole bunch of them, in fact.

Just a year after Piazzi's discovery, astronomers found a second speck in roughly the same orbit. This sister to Ceres was given the godly name Pallas, and most astronomers accepted it as a planet. But Herschel questioned whether the two newfound worlds really deserved to be put in the same category as Jupiter and the other celestial deities. By his estimate, Ceres and Pallas were so small as to be "beyond all comparison less than planets."

In a letter to a fellow member of the Royal Society, William Watson, Herschel complained that "it appears to me much more poor in language to call them planets than to call a *rasor* a *knife*, a *cleaver* a *hatchet*. . . .

"Now as we already have Planets, Comets, Satellites, pray help me to another dignified name as soon as possible," he told Watson.

Within a month, Herschel came up with another name: asteroids. The term was derived from the Greek word for star, "*astēr*," and means "starlike." Herschel made up the word because he saw the objects as starlike points of light rather than planetlike disks in his telescope.

In 2003 and 2004, the Hubble Space Telescope took the highest-resolution pictures of Ceres ever made, showing it as a round disk with a mottled, thoroughly planetlike appearance. "If Herschel had seen the disk of Ceres he might not have objected to its planetary status," said Mark Sykes, director of the Arizona-based Planetary Science Institute and a member of the scientific team behind the first NASA mission to Ceres. "If he had seen the smaller asteroids, and their irregular shapes, I suspect that he would have drawn the classical line for planet as those objects observed to have round disks. In this case, Pluto's planetary status would never have been in question."

At first, Herschel's made-up term went over about as well as Georgium Sidus. One astronomer complained about Herschel's "idle fondness for inventing names."[8] Other critics suggested that Herschel wanted to distinguish asteroids from planets just to make his own discovery seem more important.[9]

In the decades that followed, the zone between Mars and Jupiter yielded up still more mini-worlds. Their discoverers dutifully gave them mythological names as well as increasingly elaborate planetary symbols. Gradually, however, the symbols and the lists became unwieldy. Ceres and its scores of siblings were set aside in a separate category of minor planets, and numbers instead of symbols were used as shorthand. (Ceres was given the number 1.)

Eventually, the term that Herschel invented took hold after all.

Today, the International Astronomical Union's Minor Planet Center lists more than 200,000 objects, all numbered. Most of them, like Ceres, are in the main asteroid belt between Mars and Jupiter, or out beyond the planet Neptune. (Even Pluto has a number nowadays.) But some asteroids stray across planetary orbits—including Earth's. One such space rock is thought to have blasted the coast of Mexico's Yucatan Peninsula sixty-five million years ago, setting off a thermo-nuclear-scale explosion and dooming the dinosaurs.

The IAU lists so many asteroids that there aren't enough gods to go around. Finds have been named after esteemed scientists (Herschel, Einstein, and Hawking), cultural icons (Elvis, Sinatra, and Mister Rogers), places (Latvia, Las Vegas, and Bora-Bora), and even favorite pets (Petrina, Sepprina, and Mr. Spock—in this case, the discoverer's late lamented cat, not the *Star Trek* character).[10]

Asteroid discoverers have been known to take requests. For instance, I played a small role in getting a space rock named after science-fiction humorist Douglas Adams, author of *The Hitchhiker's Guide to the Galaxy* series. The asteroid I suggested for the honor had the provisional designation of 2001 DA42, which included the year of Adams's death (2001), his initials (DA), and his whimsical answer to the ultimate question (the number 42). Astronomer Brian Marsden, who headed the Minor Planet Center at the time and played a key role in the Pluto controversy, was so tickled by my suggestion that he persuaded the LINEAR asteroid search team to go along with the idea.[11]

Why are there so many space rocks in the asteroid belt? In Herschel's day, astronomers speculated that the asteroids were scattered pieces of a much larger planet that broke up ages before. Today, however, the prevailing view is that Ceres and its smaller siblings were built up through the same process that gave rise to the solar system's far bigger planets.

In the beginning, 4.5 billion years ago, the sun condensed from a huge cloud of dust and gas, with a dusty disk swirling around the infant star's midsection. Thanks to mutual gravitational and electrostatic attraction, grains of dust and ice clumped together into bigger and bigger balls of ice and dust. Over time, radiation from the sun burned off the surrounding dusty haze as if it were a morning fog, leaving behind dirty snowballs (or perhaps gassy dirtballs).

The lumpy disk thus became a snowball shooting gallery. Some of the balls became large enough to gobble up smaller ones, or slingshot them out of the solar system altogether.[12] The major planets were the big winners in the snowball fight.

The gasball now known as Jupiter was particularly prone to throw its weight around, which was bad news for the asteroid belt. Jupiter's gravitational effect stripped away space rocks, removing some of the raw material that might have clumped up into a large planet. Ceres was the biggest of the rocks left behind. Today, it accounts for as much as a third of the asteroid belt's total mass.

Was Ceres a loser in the planet-forming process, or a survivor? That's like asking whether the glass is two-thirds empty or a third full. It's a cosmic pimple even tinier than Pluto, but it's big enough to have taken on a round shape—and scientists now say it seems to have a crust, mantle, and core, just as Earth does.[13]

Scientists had no way of knowing all that in the nineteenth century, and still it took decades for the planet versus asteroid question to play itself out. In the 1850s, Ceres and a few other asteroids were included along with the major planets in the astronomical catalogs of the time, while other, smaller asteroids came to be listed in the back of the book as minor planets.

In addition to those little planets, the *Berlin Astronomical Yearbook*'s back-of-the-book list for 1854 included a mysterious new world that eventually got a promotion to match Ceres's demotion: Neptune, the planet that set the precedent for Pluto.[14]

Neptune was found because something didn't add up about Uranus. In the decades after Herschel discovered his Georgian Star, astronomers gathered more and more data about the newfound planet's orbit—including sightings that were recorded even before people realized Uranus was a planet. Using the formulas put forward by Isaac Newton, astronomers calculated how the gravitational influence of the six other planets should affect Uranus's course around the sun.

The problem was that Uranus's observed orbit didn't match up with the mathematical calculations. Either Newton was wrong, or the observations were inaccurate, or the astronomers were missing something big.

In 1821, French mathematician Alexis Bouvard suggested that the discrepancy was due to "some extraneous and unknown influence which may have acted on the planet."[15] Over the two decades that followed, more and more astronomers wondered whether the gravitational pull of yet another planet was affecting Uranus's orbit. The mystery planet's extra mass would explain the slight, puzzling changes in Uranus's orbital speed and position. If you could work out the right mathematical solution to the orbital problem, it just might point you to the mystery planet's location in the sky.

But finding a solution wouldn't be easy. You'd have to start out with a hypothetical planet, and then repeatedly fiddle with the orbit and the mass until you found a solution that fit the data. This is exactly what rival teams of planet hunters did in the mid-1840s. In France, mathematician Urbain Jean-Joseph Le Verrier publicly presented a series of reports narrowing down the range of planetary predictions, while mathematicians and astronomers in England, guided by John Couch Adams's calculations, secretly pursued their own quest.

Le Verrier tried to get French astronomers to look for the planet, but the astronomers were reluctant to invest all that time and effort in what they saw as a highly speculative and suspicious mathematical exercise. Finally, in September 1846, with the British closing in on the prize, a frustrated Le Verrier sent a letter to a German astronomer of his acquaintance, Johann Gottfried Galle at the Berlin Observatory. "I would like to find a persistent observer, who would be willing to devote some time to an examination of a part of the sky in which there may be a planet to discover," Le Verrier wrote.

Galle took the hint. Just hours after he received Le Verrier's letter, he and an astronomy student named Heinrich d'Arrest turned the observatory's telescope to the area of the sky that Le Verrier specified, and checked what they saw against a detailed star atlas. One faint point of light stood out. "That star is not on the map!" d'Arrest declared.

Two days after getting the tip, Galle sent a letter back to Le Verrier. "Sir," he wrote, "the planet whose position you have pointed out *actually exists.*"[16]

The British had been beaten, and most of the glory went not to Germany's Galle, but to France's Le Verrier. French astronomer François Arago—no impartial observer—hailed him as the first man to discover a planet with "the point of his pen."[17] Arago wanted to call the newfound world "Le Verrier," and for a while the French followed his lead. (To be consistent, they also called Uranus "Planet Herschel.")

Those names didn't stick. Le Verrier himself insisted on calling his planet Neptune, after the Roman god of the sea. And that's how it's known today. Later on, astronomers and historians determined that Galileo had observed Neptune in 1612 and 1613, but assumed it was a fixed star.[18]

The discoveries of the past four hundred years all go to show that the course of the planet debate never ran smooth, even before Pluto came onto the scene. Figuring out our complex cosmos can be a messy process, with plenty of room for miscalculation and misclassification, for clashing politics and clashing egos. Even the best of us can get it wrong, at least temporarily.

But even wrongheaded science can sometimes produce the right results—and few examples demonstrate that more clearly than the discovery of Pluto.

3

THE SEARCH FOR PLANET X

The nineteenth century spawned a long list of rich eccentrics—and if you were to alphabetize that list, you'd find Percival Lowell noted somewhere between W. K. Kellogg (cereal czar and health faddist) and Bavaria's King Ludwig II (builder of fairy-tale castles and patron of composer Richard Wagner).

Lowell possessed a rare combination of conventional connections and unconventional aspirations. On one hand, he was the scion of one of Boston's oldest, most established families. There's a whole city in

Percival Lowell seated at the Lowell Observatory's 24-inch telescope, circa 1907.

Massachusetts named after the Lowell clan. His brother was the president of Harvard, for heaven's sake.

On the other hand, Lowell was about as far removed from the staid society of nineteenth-century Boston Brahmins as you could get. At one point, he dabbled in psychic research in Japan. He took much of the fortune he amassed through his early business dealings and spent it on the astronomical venture that consumed his attention in later life: the Lowell Observatory, which was founded in 1894 and is still thriving today.

The most eccentric thing about Lowell was his campaign to convince the world that aliens were building canals on Mars, canals he thought he was uniquely suited to see. It was a campaign that would set him at odds with the scientific establishment, but would also lead by twists and turns to the discovery of Pluto fourteen years after his death.

Lowell was part of a grand generation of millionaires who backed astronomical projects in the late 1800s and early 1900s.[1] His family's fortune rested on the cotton trade, and after graduating from Harvard with distinction in mathematics, he put his skill with numbers to work in the family business. Through the years, Lowell diversified his portfolio, making shrewd investments in railroads and other utilities to build up his personal estate to more than $2 million—and that was back when a million dollars really meant something.[2]

But Percival Lowell didn't aim to make his mark as an industrialist or an investor. Instead, exploration was his passion. Starting at the age of twenty-eight, Lowell traveled extensively to the Far East. He played a diplomatic role on a Korean mission to the United States, and later studied the trance states of Shinto believers in Japan—a project that paralleled the "scientific" studies of spiritualism undertaken in the West by his contemporary William James.

Drawing upon his travels, Lowell gave lectures and wrote books about the mystic Eastern psyche, which he held to be inferior to the Western scientific mind-set. His fascination with exotic frontiers, and his high regard for the scientific frontier, hinted at the turn he took toward astronomy in the 1890s.

It all started with Italian astronomer Giovanni Schiaparelli and his observations of *canali* on Mars, features that looked like lines crisscrossing the Red Planet's disk. Lowell saw Schiaparelli as a nineteenth-century Columbus who had discovered a new civilization through the sights of his telescope, and he resolved to see the canals for himself when Mars made a close approach to Earth in 1894. Thus was a gentleman astronomer born.

Lowell scouted out a number of locations for good "seeing"—that is, the atmospheric conditions that were most conducive for telescope observations. He settled upon the dry, high, remote area around Flagstaff, Arizona, as the site for his own observatory. When Lowell peered through the telescope he borrowed from Harvard, he saw what he hoped to see: straight lines, even double sets of lines, that he held up as proof that an extraterrestrial civilization had built canals to channel Martian water.

The canals, of course, were an artifact of limited telescope technology and the almost unlimited human ability to make patterns out of fuzzy phenomena—the same sort of visual trickery that sparked all the fuss over the "Face on Mars" in more recent times. Pluto's eventual discoverer, Clyde Tombaugh, saw the canals for himself through the Lowell Observatory's 24-inch telescope when he reproduced the parameters that Lowell used years before.

"They were not figments of Lowell's imagination. I'll vouch for that; he was being honest with what he saw," Tombaugh said.[3] When he upped the power on the telescope, the seeming canals broke up into less regular patterns of dark and light.

Lowell's books and lectures about Mars's canal builders made a big splash with the general public, but they put him on a collision course with other researchers. Eminent scientists accused Lowell of bending the facts to fit his far-out preconceptions. In response, Lowell insisted that his "acute eye" allowed him to spot faint features unseen by his critics.

His scientific assistants were sometimes caught in the crossfire. One of them, A. E. Douglass, wrote a letter to the observatory's acting director, complaining that other astronomers didn't give Lowell any credit "because he devotes his energy to hunting up a few facts in support of some speculation instead of perseveringly hunting innumerable facts and then limiting himself to publishing the unavoidable conclusions, as all scientists of good standing do."

When Lowell heard about that, Douglass was promptly fired.

As the scientific establishment moved toward a consensus that Mars's canals were merely a mirage, the embattled Lowell quietly turned his attention to a campaign that he hoped would shore up his reputation: the search for a planet beyond Neptune.

The sensation over Le Verrier's discovery of Neptune in 1846 made the planet search into a cottage industry. After his Neptunian triumph, Le Verrier turned his attention to a discrepancy in the calculated orbit of Mercury. He figured that there should be yet another unseen planet, or perhaps a swarm of asteroids, circling the sun inside Mercury's orbit. Astronomers spent decades looking for the theoretical planet, which came to be called Vulcan.

"For many astronomers, Vulcan and Neptune both existed because Le Verrier's calculations demanded that they exist," Vanderbilt University astronomer David Weintraub wrote.[4] Every once in a while someone would report seeing it as a dark spot moving across the sun's disk, but they eventually concluded that these were sunspots.

The Mercury mystery wasn't truly solved until after the turn of the century, when the discrepancy was explained as a consequence of Albert Einstein's general theory of relativity. Einstein figured out that the gravitational mass of the sun was warping the fabric of space-time where Mercury made its rounds. He wrestled with the mathematics for years, and in 1915 came up with a set of equations showing that the warp factor would throw Mercury's orbit off by just the right amount.[5]

Astronomers were also looking for additional planets on the solar system's far frontier: Even when Neptune was taken into account, some astronomers still saw a discrepancy between Uranus's observed orbit and the mathematical calculations for where the planet should have been. Some of the calculations suggested that one or two extra planets might be skulking around, perhaps two or three times as far away from the sun as Neptune. William Pickering—who was one of Lowell's Harvard chums until he became a scientific rival—claimed that a complete resolution of the discrepancy required as many as seven unseen planets, known as Planets O, P, Q, R, S, T, and U.

This challenge looked appealing to Lowell, because the solution relied on cold, hard calculations and observations. If Lowell found the missing planet, which he called Planet X, all the unpleasantness over the Martian canals would give way to the sort of acclaim that Le Verrier enjoyed. So, starting in 1905, he devoted more and more of his observatory's resources to the planet search.

He turned his own mathematical skill to the problem of working out Planet X's position in the sky. Lowell figured that the planet he was looking for had to be at least as big as Earth, and more likely several times bigger. To figure out where to look, he employed as many as four "calculators"—in those days, the term meant people, not adding machines.[6] The most likely spot was determined to be somewhere in the constellation Gemini . . . or was it Libra? Over the course of a decade, Lowell pressed his assistants to make three separate photographic surveys of the Planet X hunting grounds, using progressively better equipment.

In retrospect, Lowell's search suffered from fatal flaws. He tended to concentrate on the places that were identified through mathematical calculations, and when those places didn't pan out, he'd move on to new calculations. Once again, he was looking for a few facts that would prove his case, rather than seeing the big picture.

Ironically, Pluto was captured on a couple of photographic plates that were made in 1915, but it was fainter than the kind of object that Lowell expected to see, and so it remained undiscovered.

An even bigger irony was that there was actually no discrepancy in Uranus's orbit, and thus no need for a Planet X. The discrepancy showed up in the mathematical calculations only because the estimates of Neptune's mass and position were wrong. The conclusive evidence for those past errors came in 1993 when E. Myles Standish analyzed data from NASA's Voyager mission.[7]

Even before Lowell began his search, some of the experts were beginning to suspect that Planet X wasn't really there. Lowell persevered nevertheless, hoping that his redemption would be found amid the stars.

Lowell never did find his planet. He died of a cerebral hemorrhage in 1916, reportedly just after blowing up in anger at a servant.[8] In his will, he left his personal effects, his car, some investment income, and $150,000 in cash to his widow, Constance—but he put the bulk of his multimillion-dollar fortune in a trust to support the Lowell Observatory. That didn't sit well with Constance Lowell, and her legal challenges tied up the will for a decade.[9] For all that time, the planet search had to be put on hold.

By the time the will was sorted out, hardly anyone—except perhaps for William Pickering—was still actively involved in the search for new planets. But the Lowell Observatory's director, Vesto M. Slipher, was at last free to resume the quest that his late boss started. The observatory purchased a brand-new 13-inch astrograph, a telescope equipped with a camera that could spot objects much fainter than anything Lowell could have seen.

Slipher and the observatory's trustee, Roger Lowell Putnam, decided they needed to hire someone to help out with the search. Someone who didn't already have a research agenda of his own. Someone, perhaps, like the "young man from Kansas" who had sent samples of his astronomical drawings to Slipher in hopes of getting some pointers.

So it was that twenty-two-year-old Clyde Tombaugh, who worked on the family farm in western Kansas by day and gazed through his homemade 9-inch telescope by night, came to the Lowell Observatory in late 1928. Tombaugh was born on a farm near Streator, Illinois, a year after Lowell began the search for Planet X. He was a ten-year-old stargazer when Lowell died in 1916. He moved west to Kansas with his family in 1922, while Lowell's widow was wrangling over her late husband's will. Now Tombaugh was being groomed to continue Lowell's legacy.

Like everyone else around the observatory, Tombaugh was expected to help out with the chores: giving tours, stoking the furnace, even climbing up onto the canvas-covered telescope dome and shoveling off the snow in the wintertime. But his primary job was to stay up during the night and take a series of one-hour photographic exposures, capturing patches of the star field on 14-by-17-inch glass plates.

The same patch of sky would have to be photographed at different times—and that's when the real work began. The point of the exercise was to compare the plates meticulously, in hopes of finding spots that changed their position over time. By precisely measuring the distance between the two images of a moving object, astronomers could calculate how far away the object's orbit was.

A fast-moving object would be closer than a slow-moving object. The closer spots were most likely asteroids or comets, and the observatory's astronomers saw a few of those. But what they were really looking for was an object moving slowly enough to demonstrate that it was beyond the orbit of Neptune.

Decades earlier, when Lowell started his search, he would lay two glass plates on top of each other, slightly offset, and then peer through a magnifying glass looking for the spots that moved. Then, in the latter years of his search, the observatory acquired a device known as a blink comparator, a setup that moved a microscope apparatus quickly back and forth between one plate and another—*clack, clack, clack*—for a visual comparison of the star field.

When the search resumed in April 1929, the observatory's senior researchers were supposed to take turns "blinking" the sets of photographic plates that Tombaugh made. But the researchers became distracted by other duties, and the plates started to pile up. In June, less than a year after Tombaugh arrived in Arizona, Slipher told the Kansas farm boy that he would be in charge of blinking the plates as well as making them.

Clyde Tombaugh peers into a blink comparator at the Lowell
Observatory, circa 1950. Such devices were used to check photographic
plates for tiny, faint objects that moved between one exposure and
another—objects such as Pluto.

"I was overwhelmed," Tombaugh recalled. "It had become
evident to me that the one doing the blinking carried the
heavy responsibility of finding, or not finding, the planet."[10]

The hunt for Planet X thus became something radically
different from the hunt for Neptune. For Le Verrier, the main
task was to refine the mathematical calculations, time after
time, until he came up with a solution that fit the data so
precisely he could tell any astronomer willing to listen where

to look. Lowell tried the same strategy, but the observations stubbornly refused to conform to the calculations.

By necessity, Tombaugh's approach was a throwback to the good old days of Herschel's time: It turned into a wider, more painstaking survey of the ecliptic, the equatorial zone in the night sky where planets wandered. "The project in a sense passed out of the mathematics of Percival Lowell and into the observational technique of Clyde Tombaugh," astronomer David Levy, codiscoverer of a famous comet that smacked Jupiter in 1994, wrote in his biography of Tombaugh.[11]

Today, astronomers are finding Pluto's kin on the edge of the solar system using the same basic technique, supercharged with automated telescopes, high-powered cameras, and sophisticated software to sift through gigabytes' worth of imagery. But in 1930, the telescope, the camera, and the analysis were all guided by one young man: a country boy who went out west because he didn't want to go into farming and couldn't afford to attend college.

Fortunately, this country boy was a dogged observer, unlike Lowell. No one could accuse him of failing to hunt down "innumerable facts." He fell into a workday routine that was as grueling as a Kansas farmer's schedule.

On a typical day, Tombaugh would get up in the morning, eat a light meal, and check the weather. If the skies looked favorable, he'd make his plans for a night of observations. Some of the day might be spent doing the chores, or developing plates, but the real workday began after dark. Sitting beside the Lowell Observatory's 13-inch triplet-lens telescope, Tombaugh would take pictures of the area of the night sky directly opposite the sun. If he was lucky, he could get in several hourlong plate exposures during the course of the night, and then fall into bed before dawn at the end of a fourteen-hour workday.

If clouds looked likely to spoil the night's observations, or if the moon was due to interfere, Tombaugh would spend the day blinking plates. This involved sitting down at the comparator, looking through the microscope eyepiece, and checking every square inch of star-filled photographic plates. The task required so much concentration that Tombaugh had to take a break every half hour. During the breaks, he might write down lab notes about what he saw, have lunch, read a journal, or chat with the other astronomers about Planet X, Mars, or other matters.

On the morning of February 18, 1930, Tombaugh could tell it was a day for blinking. The skies over the observatory were cloudy—and the glare from a last-quarter moon would probably spoil the picture-taking even if the skies cleared up. He hunkered down for what he reckoned would be a nine-hour day of checking photographic plates.

As it happened, the plates he selected had been made the previous month, when the telescope was pointed at an area of the sky that Lowell targeted all those years before. The familiar clack of the comparator sounded as Tombaugh slowly made his way across the surface of the plate. He put in several hours of blinking, broken up by short breaks at the office and a lunch hour at a downtown café.

By four o'clock, Tombaugh had examined about a quarter of the plate, but there was nothing to see except for faint flaws in the photographic emulsion. He clicked past Delta Geminorum, a bright star about six degrees away from a point Lowell had once favored for the location of Planet X. Then he saw it.

As the comparator clicked, a dark spot popped in and out of his field of view. When Tombaugh moved the microscope a bit to the side, another spot appeared and disappeared—*clack, clack, clack.* He turned off the automatic blinking. The spots were several times fainter than Lowell had predicted for Planet X, but they were definitely there. And the object recorded by those spots was definitely moving in the right direction.

How far away was the object? Tombaugh took a plastic ruler and measured how far apart the spots were. The distance was 3.5 millimeters, or about an eighth of an inch. That meant the object couldn't be a close-in asteroid. If the object was genuine, it had to be farther away than Neptune.

For the next forty-five minutes, Tombaugh's mind raced through a flurry of questions: Could the spots be just a fluke? Were they just smudges on the glass? Or maybe they were two variable stars that just happened to flare close together? To answer those questions, he pulled out a third photographic plate from January and took a close look. Sure enough, there was a faint extra spot just where he expected it to be.

"I was walking on the ceiling," Tombaugh recalled later. "I was now 100 percent certain."

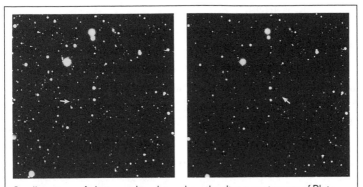

Small sections of photographic plates show the discovery images of Pluto moving across the night sky. Pluto is indicated by a white arrow superimposed on each plate. The left plate is from January 23, 1930. The right is from January 29, 1930. Lowell Observatory astronomer Clyde Tombaugh found Pluto as he examined these plates on February 18, 1930.

Clyde Tombaugh thus became the first man in history to take notice of little Pluto. After checking a few more plates, just to make sure, it was time to tell the world. Tombaugh walked across the hall to astronomer Carl Lampland's office and filled him in. Lampland immediately ran over to look at the plates through the comparator. Meanwhile, Tombaugh continued down the building's long hallway and walked into Slipher's office.

"Dr. Slipher," Tombaugh announced, "I have found your Planet X."[12]

It has often been said that if Pluto were discovered today, it would not have been considered a planet, due to its lack of mass or breadth or orbital dominance. But it doesn't seem right to gauge Pluto's status—or the status of anyplace else, for that matter—without taking the historical circumstances into account. How would the Tigris River rank among waterways if it weren't part of the cradle of civilization? How would Mount Sinai rate as a mountain if it weren't mentioned in the Bible?[13]

The tale of Pluto's discovery has history galore. The search for Planet X took longer than the search for Neptune, and eventually thrust an unlikely hero from Kansas into the scientific spotlight. The strangest part of the story is that the highly eccentric world was found because of a highly eccentric millionaire's mistake. It turned out to be pure coincidence that Pluto was found in the same part of the sky that Lowell started searching twenty-five years earlier. Lowell might never have started that search if he had had the right numbers for the orbits and the masses of Uranus and Neptune.

But if Lowell had not embarked on his quest for redemption, Pluto might have gone unnoticed for years longer. And who knows? The tale of the glittering realm on the edge of our solar system might have spun off in a completely different direction.

4

PLUTO AND ITS LITTLE PALS

As the first new world discovered in eighty-four years, Planet X was primed to cause a sensation. Tombaugh and the rest of the Lowell Observatory team kept the news under wraps until March 13, 1930, which marked Percival Lowell's birthday as well as the 149th anniversary of Herschel's discovery of Uranus. In the meantime, the astronomers checked and rechecked their find, on the photographic plates and through the telescope.

One question nagged at Tombaugh during the weeks leading up to the scheduled announcement: If this was Planet X, why was it so small? Lowell's calculations had predicted that Planet X would be as much as seven times as massive as Earth—and it should have shown up as a bright disk. This object, however, was visible only as a point of light.[1]

At one point, Tombaugh worried that he had merely identified the moon of a larger planet yet to be found. Perhaps the true Planet X was lurking elsewhere among the observatory's stacks of photographic plates. "That just scared the living wits out of me for a while," he said.[2] He was relieved to hear from his colleagues that they couldn't find any other planets on the plates.

Many of the questions about Tombaugh's Planet X were still unresolved when the appointed day came. Shortly after midnight Eastern time on March 13 (when it was still March 12 in Flagstaff), Vesto M. Slipher sent a telegram announcing the find to the Harvard College Observatory, which served as the Western Hemisphere's clearinghouse for astronomical discoveries. Harlow Shapley, the director of the Harvard observatory, immediately published the announcement and forwarded it to the International Astronomical Union's Central Bureau for Astronomical Telegrams, which in turn distributed the news around the world.

In a follow-up circular, Slipher acknowledged that the object was being publicized "before its status is fully demonstrated, yet it has appeared a clear duty to science to make its existence known in time to permit other astronomers to observe it while in favorable position." The wording of the circular handled the question of planethood delicately, noting only that it seemed to match Lowell's predictions for "a planet beyond Neptune."[3]

The press reports were far less circumspect. On the day after the observatory's announcement, the *New York Times* trumpeted the news at the top of its front page: "Ninth Planet Discovered on Edge of Solar System; First Found in 84 Years."

Those first reports downplayed Tombaugh's role as the discoverer. For example, in the Associated Press dispatch picked up by the *Times* and many other papers across the country, Tombaugh's name didn't appear until the ninth paragraph. Slipher and eight other astronomers were mentioned before him. What's more, Tombaugh was credited not as an astronomer per se, but as a "photographer at the observatory, who saw a tiny spot on one of his plates."[4]

A century and a half earlier, King George III had made Herschel his private astronomer to reward him for finding a planet. Tombaugh's reward was more modest: The University of Kansas gave him a four-year scholarship, and the onetime farm boy could at last afford to get his college degree.

The times were tailor-made for fascination with Planet X. The Great Depression was just taking hold after the stock market crash of the previous October, and far-out flights of fancy provided a welcome diversion from the gloom. Science-fiction adventures were standard fare in the pulp magazines of those days. *Buck Rogers in the 25th Century A.D.* made its comic-strip debut just a year before Planet X's discovery.

Now that Planet X had made its own kind of debut, it was up to the discoverers to give it a proper name. That job wasn't as easy as it might have sounded, considering the precedents set by the Medicean Stars, Georgium Sidus, Ceres Ferdinandea, Planet Herschel, and Le Verrier's Planet. Many names were suggested, by members of the Lowell Observatory as well as the public, and eventually the list was whittled down to three: Minerva, the Roman name for the goddess of wisdom (Athena to the Greeks); Cronus, the son of Uranus and the ruler of the Titans in Greek mythology; and Pluto, the Roman god of the underworld.

At first, Minerva was the top choice—but then astronomers learned that an asteroid had already been named after that particular goddess. Slipher ruled out Cronus because a "certain detested egocentric astronomer" had been using that name for his own hypothetical ninth planet. Slipher feared the astronomer might claim a share in the glory if the planet was called Cronus. (The astronomer in question was Thomas Jefferson Jackson See, a former member of the Lowell Observatory's staff.)[5]

Pluto *might* work. It seemed fitting to name the coldest, dimmest, most remote planet after the dark lord of the dead. But the observatory's trustee, Roger Lowell Putnam, worried that the name would be associated instead with Pluto Water, a sulfurous mineral water that was bottled at Pluto Springs in

Indiana and marketed as a fast-acting laxative. "When nature won't, Pluto will" was a popular ad slogan at the time.[6]

Counterbalancing the association with constipation was the fact that the scientific symbol formed from the first two letters of the planet's name—P-L—would evoke Percival Lowell's memory. That just might mollify Lowell's widow, Constance, who insisted that the planet should be called Zeus, or perhaps Percival, or even Constance.

And when it came to public relations, Pluto had another big advantage: The suggestion could be credited to an eleven-year-old girl from England named Venetia Burney.[7]

Seventy-five years later, Venetia Burney Phair recalled having breakfast with her family on the day after the planet's discovery was announced. "My grandfather read out at breakfast the great news and said he wondered what it would be called," she told an interviewer. "And for some reason, I—after a short pause—said, 'Why not call it Pluto?' I did know, I was fairly familiar with Greek and Roman legends from various children's books that I had read, and of course I did know a little bit about the solar system and the names the other planets have. And so I suppose I just thought that this was a name that hadn't been used. And there it was. The rest was entirely my grandfather's work."[8]

Little Venetia was lucky to have the right grandfather for the job: Falconer Madan had been the head of Oxford's renowned Bodleian Library. What's more, planet-naming ran in the family. His older brother Henry was the one who suggested naming the moons of Mars after Phobos and Deimos, the Greek gods of fear and terror.

So right after breakfast, Madan passed along Venetia's suggestion in a note to Oxford professor Herbert Hall Turner, who had once been Britain's astronomer royal. Turner got Madan's message after returning from a Royal Astronomical Society meeting where Planet X's discovery was topic A. The eminent astronomer was impressed with the girl's suggestion. He quickly sent a telegram about it to the Lowell Observatory, and told Madan that "Miss Venetia will get the best chance I can give her."[9]

On May 1, Slipher announced that Pluto was the Lowell Observatory's choice, and that Venetia had provided the inspiration. To celebrate, Madan gave his precocious granddaughter a five-pound note—a reward she still remembered decades later. "This was unheard of then," she said. "As a grandfather, he liked to have an excuse for generosity."[10]

Not everyone was thrilled with the way the naming process turned out, however. William Pickering, the rival astronomer who had predicted the theoretical paths of Planets O through U, complained that he was planning to use the name Pluto once Planet P was found. "Pluto should be named Loki, the god of thieves," he grumbled.[11]

There was also the fact that Planet X was discovered in the same area of the sky where Pickering had placed his own Planet O—leading some to wonder whether the newfound planet was predicted not by Lowell but by Pickering instead.

Eventually, Pickering reconciled himself to the choice of Pluto and its P-L symbol. "That's a good name—Pickering-Lowell," he once told a visitor.[12]

.   .   .

After being named by a youngster, Pluto was quickly clasped to the hearts of youngsters—with the help of an up-and-coming cartoon animator named Walt Disney. The dog that came to be known as Pluto made his first appearance in a Disney movie in October 1930, shortly after Planet X got its name. Initially, the bloodhound character was Minnie Mouse's pet, called Rover. But that name lasted about as long as Georgium Sidus, Herschel's chosen name for Uranus. The pup popped up again the following May in another animated short, "The Moose Hunt," with a new owner (Mickey Mouse) and a new name.

A Disney news release, framed as a quotation from Mickey, made it sound as if Walt Disney picked the name Pluto out of thin air: "Names like Rover and Pal were dreamed up, but none seemed to fit. One day Walt came in and said, 'How about Pluto the Pup?'—and Pluto it's been ever since."[13] But the timing of Pluto's renaming led to the suspicion—even among Disney historians—that the great animator decided to capitalize on the astronomical sensation of the day.[14]

The new planet Pluto was all the more appealing to youngsters because it was small like them. To grown-up astronomers, however, that smallness was a source of puzzlement. After all, Lowell had begun looking for Planet X because he believed there must be something out there big enough to affect the orbits of Uranus and Neptune. But as astronomers pieced together additional data about Pluto and the closer planets, they confirmed that Pluto was too small to have any gravitational effect.

Then there was the fact that no one could make out Pluto's disk, even through the most powerful telescopes of the time. That left no direct way to measure how wide it was. Instead, astronomers tried an indirect route to estimate the planet's size, based on mathematical calculations that took three factors into account:

- Pluto's distance, which they could calculate from changes in the planet's position over time. Piecing together all the sightings that had been made, astronomers figured out that Pluto orbited the sun every 248.54 Earth years at a distance ranging from 2.7 billion to 4.7 billion miles. That's equivalent to 30 to 50 AU (astronomical units), where 1 AU equals the distance from Earth to the sun, or 93 million miles.[15]
- Pluto's brightness as seen from Earth. The planet was the equivalent of a magnitude-15 star, which was close to the limit of what the Lowell Observatory's 13-inch telescope could detect.
- The reflectivity of Pluto's surface. That was the squishiest number in the equation, but astronomers could come up with some ballpark figures based on how reflective different substances could be.

When astronomers ran the equations, they found that Pluto couldn't possibly be several times bigger than Earth. Under the most optimistic assumptions, it might approach Earth's size if its surface was rocky and dull. But if the surface was icy and shiny, Pluto was just a fraction of our planet's size.

For the better part of five decades, astronomers labored to determine just how small Pluto was, based on guesses about its composition. Every fresh estimate downsized it a little bit more—leading some scientists to joke that the planet would eventually shrink to nothingness.

The Lowell Observatory's Carl Lampland—Tombaugh's office neighbor—took the first crack at figuring out what Pluto was made of, just days after the planet was discovered. Lampland looked at the point of Pluto's light through different-colored filters, and determined that the hue was much yellower than Neptune.

Under the right circumstances, a planet's color can reveal an amazing amount about its composition. What you have to do is obtain a spectrum for the light reflected by the planet—a breakdown of the light that shows precisely which wavelengths are being reflected, and which are not. Every element has a characteristic spectral signature, and if your spectrograph is good enough to separate out those signatures, you can figure out which elements are present in which proportions.

Unfortunately, the spectrographs weren't nearly good enough in those days to analyze the faint light reflected by Pluto. Instead, astronomers conducted simpler experiments to measure how the brightness of the light varied over time. Even those limited experiments provided fresh clues to Pluto's puzzles.

For instance, Vanderbilt University's Bob Hardie and Merle Walker noticed in the 1950s that the planet went through a cycle of brightening and fading every 6.387 days. The pattern was as regular as clockwork. That told astronomers that Pluto must be making one complete turn on its axis in that amount of time. Just by measuring a point of light on a regular basis, scientists could figure out the length of one Plutonian day.

Finally, in 1970, astronomers from the University of Iowa recorded the first published spectrum of Pluto.[16] Over the years that followed, the spectral readings became better and better. Astronomers eventually concluded that Pluto's surface contained frozen nitrogen as well as bright methane ice, which would reflect as much light as fresh-fallen snow on Earth.

If a celestial body's surface is more reflective than expected, then it would require less surface area to shine with a given brightness. The fact that methane frost was detected on Pluto's surface meant the size estimates for the incredible shrinking planet would have to be shrunk even more.

By the late 1970s, Pluto's diameter was estimated at somewhere between 1,000 and 3,300 miles. At best, Pluto might be as big as Mercury; but at worst, it was smaller than Earth's moon. How much smaller? In the summer of 1978, an astronomer made a discovery that would provide the key to answering that question.

5

THE MEANING
OF A MOON

For almost fifty years, finding out anything at all about Pluto was devilishly difficult, but finding out that it had a moon took just two days.

That's how much time astronomer James Christy set aside at the U.S. Naval Observatory in Washington to look through photographic plates showing Pluto's position among the stars at various times. The plates had been made at the Naval Observatory's astronomical facility in Flagstaff, not far from the Lowell Observatory.

The pictures of Pluto on some of the plates seemed to be oddly distorted. Perhaps the plates were defective. Maybe the Naval Observatory's 61-inch telescope had failed to track the stars properly, or maybe it was just that the "seeing" wasn't good on those particular nights. Christy had developed an expert eye in the course of analyzing tens of thousands of plates for the observatory's extensive survey of double stars, and so he brought out his microscope to double-check these troublesome Pluto plates.[1]

Sure enough, Christy saw that ten of the pictures of Pluto were slightly elongated. But then he noticed something else: The images of the surrounding stars were perfect points. The misshapen views of Pluto were revealing something real about the planet. Moreover, when Christy compared the plates from different times, that "something" seemed to be moving from one side of Pluto to the other.

At first Christy wondered whether the elongation could be a mountain sticking up from the surface, but no mountain could be that tall. Could it be the eruption of a volcano into space? He dropped that idea as well: An eruption that big, from a planet that small, couldn't possibly last a month. Then Christy seized on the right answer. "What?" he thought to himself. "Pluto has a moon?"

The next morning, he dug into the files and checked images of Pluto going back to 1965. The pictures, and the calculations made by Christy and his colleagues, confirmed that the elongation rounded the planet every 6.387 days. The best explanation for the observations was a moon that was gravitationally locked in orbit with Pluto.

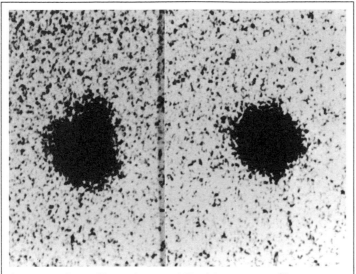

Astronomer James Christy discovered Pluto's largest moon, Charon, in 1978 by comparing these fuzzy images. Christy noticed that some images were elongated, like the one at left, while others were not. After close examination, Christy concluded that the elongations were caused by the presence of a previously unknown moon orbiting Pluto.

It took years longer to make all the observations required to confirm what Christy saw that first day, but in the end, Pluto was recognized as having something that two bigger planets in the solar system—Mercury and Venus—did not: an orbital companion.

Why did it take so long to figure out Pluto had a moon? One reason was that the photographic exposure settings had to be just right to bring out the elongation in Pluto's image. Pluto was gradually coming closer to Earth, and thus the images taken in the 1970s were sharper than the images from the 1930s or 1950s. But the biggest factor behind the find was Christy's ability to see a discovery where others just saw defects.

Once the moon's existence was confirmed to the International Astronomical Union's satisfaction, Christy was given the honor of naming it. At first he told his wife, Charlene, that he'd name it after her: It would be called "Char-on," like a proton or neutron. Then he realized that the name had to follow the Roman or Greek god pattern in order to conform to the IAU's rules. He picked up a dictionary, flipped to the *Ch-* section, and started looking for mythological names. There, to his amazement, he found a reference to the ferryman of Greek lore whose boat carried the dead across the River Styx to Pluto's dark realm: Charon!

The story sounds too good to be true, but in any case, Christy found a way to keep his wife as well as the IAU happy. Classical scholars may pronounce the mythological ferryman's name as "*Care*-on," but in accordance with Christy's wishes, most astronomers today call Pluto's moon "*Shar*-on."[2]

Every once in a while, a puzzle fan experiences a dazzling moment when the addition of one key jigsaw piece, or one crossword entry, opens up a cascade of opportunities for solving the puzzle. That was the kind of moment that astronomers experienced once they learned about Pluto's moon.

Having two data points was the key to working out the mass and the motions of two worlds that could just barely be seen as more than a single point. "Within a week of Charon's discovery there were roughly a dozen major conclusions made concerning the true nature of Pluto," Christy recalled.[3]

The angle of the elongations in the pictures of Pluto told Christy and his colleagues that the moon was tracing a nearly north-to-south orbit—a cockeyed circuit that had been seen at only one other celestial location: Uranus, the planet found by William Herschel almost two centuries earlier.

Meanwhile, the length of the elongations revealed how far away Charon was in its orbit. Assuming that Pluto was much more massive than Charon, astronomers could figure out Pluto's mass, based on equations that factored in the time it took Charon to circle Pluto (6.387 days) as well as the distance between them (12,000 miles).

The answer was shockingly small: Pluto was only 0.2 percent as massive as Earth, or about one-sixth as massive as Earth's moon. If Lowell's ghost still haunted anyone wondering whether Pluto could affect the big planets' orbits, Charon exorcised it.

To make the maximum use of their observations, astronomers employed computing power far more advanced than the four human "calculators" Lowell had hired to search for Planet X. But finding the best ways to mine the data required scientists who were extraordinarily savvy—or extraordinarily lucky. Swedish-born astronomer Leif Andersson was both.

Based on the rough readings of Charon's orbit, Andersson divined that Pluto and Charon should go through a series of mutual occultations, during which they would repeatedly pass in front of each other. By analyzing how the light dipped and flared over time—the detailed light curve for the two paired worlds—astronomers could compare their sizes and their compositions.

Pluto makes a close approach to the sun only once in the course of its 248-year orbit, and the season for seeing Pluto-Charon occultations comes only once every 124 years. As luck would have it, the prime-time season was just about to begin, although astronomers didn't know the orbits quite well enough to determine exactly when.

So the world's astronomers kept watch for the telltale pattern of dimming and brightening that would signal the start of the cosmic pas de deux. At last, in 1985, the dance began. In the beginning, each icy world cast a shadow just slightly grazing the edge of its orbital partner. As the dance continued, each eclipse covered more of each disk: Charon darkening Pluto, then Pluto darkening Charon. The shadows reached their deepest when Charon's orbit took it directly over Pluto's disk. After the climax, the orbital angle widened and the shadows ebbed.

The shadow dance took six years to go from start to finish, and during all that time, the light curve served as a cosmic CAT scan, tracing the outlines of the planet and its moon.

Pluto and Charon turned out to be the most evenly matched planet and moon known in the solar system: Charon was one-seventh as massive as Pluto, and a little more than half its diameter. If you were standing on Pluto, Charon would loom above you, looking seven times wider than Earth's moon as seen from our home planet.

Like our own moon, Charon always turns the same face toward Pluto. But unlike our moon, Charon holds a fixed place in the Plutonian heavens. Viewed from one half of Pluto, Charon would stand still while the sun and other stars whirl dimly beyond. Viewed from the other half, the moon would be perpetually missing from the skies.

The occultations were good for much more than orbital mechanics. By analyzing the changing spectrum of the light from Pluto and Charon, scientists could improve upon Lampland's observation back in 1930 that Pluto's hue was "yellowish." They could even tease out the differences in the elemental composition of the two worlds. The spectral data showed that Charon's surface was dominated by grayish water ice rather than the yellowish methane ice seen on Pluto.

One of the most amazing computational feats involved analyzing light-curve data going all the way back to 1954, and matching those readings against computer models for the shadings of Pluto and Charon. Two teams of researchers came up with "maps" of Pluto based on the models that fit the data best, and although there were some differences, there were amazing similarities as well: Both maps showed a bright, methane-rich south polar cap, and a dirty, dark region around Pluto's midriff.

Those bright spots of methane frost supported the idea that Pluto had an atmosphere that came and went with the planet's seasons: The frozen methane would break down over time, chemically changed by ultraviolet radiation from the far-off sun. Astronomers even saw evidence that some of the frost was darkening, which gave Pluto its yellowish-brownish hues. If there was fresh frost, the likeliest explanation was that it froze out of the atmosphere. Putting all the evidence together, scientists concluded that Pluto's atmospheric methane turned to frost during the winter, and then rose back into the air during the summer.[4]

Knowing the sizes and the masses of Pluto and Charon gave scientists an opening to calculate their density and guess at what kind of stuff might lie beneath the surface. Pluto's density was somewhere around 2 grams per cubic centimeter—midway between rock and water ice. That meant Pluto was no mere ball of ice, but more likely a mixture of 30 percent ice and 65 percent light minerals by mass, with traces of heavier minerals at its core. Charon's density was less than Pluto's, at about 1.2 grams per cubic centimeter. That implied that the moon was significantly icier.

Thanks to the discovery of Charon, more and more of Pluto's puzzle pieces were being fitted into place. Astronomers no longer saw Pluto as just a single point of light, but as an actual world spinning on its poles in the oddest way, with an actual terrain that could be mapped in light and dark, possessing an atmosphere that froze and thawed with the seasons, traveling in tandem with a moon hanging in the black sky.

Pluto was no longer alone on the solar system's edge. It had a partner. But that drew attention to a huge gap in the jigsaw puzzle: How did Charon get there? And were there still more icy worlds out there, waiting to be discovered?

6

THERE GOES THE NEIGHBORHOOD

Golden anniversaries are traditionally a time for celebration—but for astronomer Brian Marsden, the golden anniversary of Pluto's discovery was a reminder of how much of a misfit the planet had become.

Marsden, a British-born expert on celestial mechanics, was head of the Central Bureau for Astronomical Telegrams, the same office that had sent out the news of Pluto's discovery fifty years earlier. Since then, the bureau had been relocated from Copenhagen to Percival

Lowell's old stomping grounds in Cambridge, Massachusetts, where Marsden was the primary gatekeeper for astronomical bulletins as well as the head of the International Astronomical Union's Minor Planet Center.

It was Marsden's job to help keep track of asteroids, comets, and natural satellites—in fact, anything that popped up in someone's telescope that didn't happen to be a major planet. Pluto was one of the nine major planets, and thus not part of Marsden's domain. That was a situation he wanted to change.

Marsden got his chance when Clyde Tombaugh invited him to deliver a talk during a gala symposium on "the Ninth Planet's Golden Year" at New Mexico State University on February 18, 1980—exactly fifty years after Tombaugh spotted Pluto.

When it was Marsden's turn to speak, he started out by tracing the 199-year history of planet-hunting since Herschel's day. Then he noted that astronomers were starting to find a few asteroids that, like Pluto, crossed the orbits of giant planets. One of them, called Chiron, was even trumpeted for a while as the "tenth planet." Eventually, however, Chiron was pigeonholed as an unusual kind of asteroid that happened to cross the orbits of Saturn and Uranus. In the end, astronomers classified it as a comet as well as an asteroid because it sometimes sprouted a glowing tail. Yet another asteroid, named Hidalgo, appeared to be a burned-out comet, tracing a highly eccentric, highly inclined orbit that took it close to Saturn.

Marsden suspected that there were still more orbit-crossing oddities out there, just like Chiron and Hidalgo. And like Pluto.[1] "Is it therefore perhaps not time we dropped the appellation 'ninth planet' and *classified* Pluto with the two objects it most obviously resembles, as an unusual minor planet?" he asked the audience.[2] He said he could even give Pluto a new identity as No. 330 on his list of minor planets—replacing Adalberta, a reported asteroid that had turned out to be a celestial will-o'-the-wisp.[3]

Marsden's suggestion was couched in the polite qualifiers and wit that reflected his British upbringing. "It was partly in jest that I did this at that time," he recalled years later. But to some of his listeners, it was as if someone had stood up at a couple's golden anniversary party and announced that the wedding was a sham. At least that's how Tombaugh felt.

"My dad was crushed," said his daughter, Annette Tombaugh-Sitze. "He was very angry, and he was crushed that Brian picked this time to bring it up."

Marsden said he meant no disrespect. "I never intended to be unkind to Clyde," he insisted. Yes, he was aware that members of Tombaugh's family were still angry with him, but he said any ill will was the result of a misunderstanding. "They're not astronomers," Marsden said. "They don't see it in quite the right way."

Marsden was of the opinion that Pluto should never have been designated a planet in the first place. He said the Lowell Observatory "bamboozled" the world into thinking Pluto was the giant Planet X that Percival Lowell had predicted. But

Brian Marsden, who headed the IAU's Minor Planet Center, looks through papers at the IAU's General Assembly in Prague in 2006.

history showed that Lowell was totally wrong. In Marsden's words, he was "very much a member of the minor leagues."

And the way Marsden saw it, Pluto belonged in the minor leagues as well.

Some scientists might not have cared that much which league Pluto was in. But this was a matter that nagged at

Marsden, the man responsible for cataloging the miscellaneous bits of the solar system. His suggestion to classify Pluto as a minor planet, as a far-out kind of asteroid, was aimed at putting a misfit planet in its rightful place at last.

In the beginning, even Tombaugh had his doubts about Pluto—and there was nothing unusual in that. After all, William Herschel had to be convinced that what he saw back in 1781 was really a planet and not just a comet.

The biggest knock against Pluto was that it was so small. It takes about 25 Plutos to equal the mass of Mercury, the next largest planet on the solar system's scale, and 476 Plutos to match Earth's mass. But if you judge by volume rather than mass, the discrepancy isn't as great—largely because ice-covered Pluto is less dense than those rocky terrestrial planets. It takes just 8 Plutos to fill up Mercury's volume, and about 150 Plutos to equal Earth. In comparison, it takes 57 Earths to equal the volume of the next planet up on the size scale, Neptune.

When astronomers looked toward the other side of the size spectrum, as it was known in 1980, they saw no asteroid or comet that was nearly as big as Pluto: It's 14 times as massive as the biggest asteroid, Ceres. Even if you took Ceres and added in all the other asteroids in the main belt, you'd still have less than a quarter of Pluto's mass. And on the volume scale, it would take about 14 Ceres-sized objects to equal one Pluto.

Another knock against Pluto was its inclined orbit, which is at a 17-degree slant from the solar system's main plane.[4] But that tilt actually works in Pluto's favor: Its orbit is angled in such a way that Pluto comes closer to the sun than Neptune for twenty years at a time, while never cutting through the bigger planet's orbital path.

That's one way in which Pluto's orbit is calibrated to keep the little guy out of the big guy's way. Another has to do with the clockwork of the two planets' orbital motions. For every three orbits that Neptune makes, Pluto makes almost precisely two. When Pluto nears the track that Neptune follows, Neptune is at least a quarter of an orbit away. And when the speedier Neptune overtakes Pluto, their separate tracks are so far apart that the gravitational effect is minimal. What little effect there is serves to correct variations in Pluto's orbit, keeping it close to the two-to-three resonance. In effect, Neptune keeps Pluto in a protected zone.

The result of all this is that Pluto never comes any closer to Neptune than 17 AU. That's about 1.6 billion miles, or roughly the distance from the sun to Uranus. Pluto actually comes closer to Uranus (11 AU) than to Neptune in the course of its celestial travels.[5]

The Pluto-Neptune clockwork completes one grand cycle every 496 years or so—two full Pluto orbits, or three Neptune orbits. The ever-so-slight gravitational shifts oscillate over a far longer cycle: about 300 full orbits of Pluto, or 70,000 years.[6] This finely tuned timetable explains why Pluto has been able to stay the course for billions of years.[7]

The case for Pluto looked stronger once astronomers worked out all the implications of Charon's discovery. Here was a planet in an odd but stable orbit, with a moon and an atmosphere and surface variation. Marsden might have wanted to lump Pluto in with asteroids or comets, but when he spoke out in the 1980s, Pluto's qualities and its heft put it in a class by itself. The loneliness of the planet and its overgrown moon seemed to work in its favor.

Nagging questions remained, however. As scientists continued to study Pluto and Charon, they began to ask themselves how those two misfits ever came together. They considered a variety of scenarios—including the idea that they sprang from the same knot of gas and dust, or that Charon was a passing globe of ice drawn into its orbit by Pluto's gravitational pull. Based on all the evidence—including the differences in the makeup of Pluto and Charon as well as the characteristics of their orbits—the most likely scenario suggested that a celestial interloper slammed into Pluto billions of years ago. Pluto eventually recovered from the blow, and the lighter debris from the blast coalesced to form Charon.

Such a scenario isn't as crazy as it might sound: Our own moon is thought to have formed in the same way, as the result of a collision between a hot, infant Earth and a Mars-sized planet that got in its way. The difference, however, was that ancient Earth's cataclysm occurred in a place where planetesimals got in each other's way like bumper cars at a carnival. What were the odds that two solitary worlds—Pluto and the object it ran into—would cross paths on the solar system's very edge? When astronomers ran the numbers, the chances against that happening turned out to be . . . well, astronomical.

In order to provide a realistic probability for the kind of smash-up that gave Pluto its moon, there had to be thousands upon thousands of objects lurking beyond Pluto and the seeming boundary of the solar system. The way that Pluto and Charon were thought to have come together added to the growing suspicion that the edge of the solar system was once teeming with icy objects. If such worlds still survived, they were apparently beyond the power of telescopes to see. But not for long.

When Marsden spoke in 1980, astronomers couldn't see anything out there that came anywhere close to Pluto and Charon.

Long after Tombaugh found Pluto and got his college degrees, he kept on with his methodical search of the skies from the Lowell Observatory. He toiled over the observatory's comparator, blinking photographic plates until 1943. Over all that time he identified 3,969 asteroids, two comets, and a nova. But he also determined that there were no more Planet Xs to be found, at least within the observing capability of the telescopes at Lowell.

Was there anything at all beyond Pluto? Of course: The ancient Greeks counted comets among the wanderers of the sky, and by the time Tombaugh found Pluto, astronomers understood that the orbits of many comets swung far out-side the orbit of Neptune. Marsden even suspected that Pluto might be a comet stuck in the deep freeze, occupying an orbit that never took it close enough to the sun for a tail to flare.

The most famous comets are the ones that have been seen coming back around on a regular schedule, such as Halley's Comet. Such objects are known as short-period comets, and in 1951 Dutch-American astronomer Gerard Kuiper pro-posed that the solar system's gravitational effects pried loose short-period comets from an icy disk beyond Pluto's orbit. This hypothetical disk came to be known as the Kuiper Belt.

Then there are the long-period comets, whose orbits extend so far out that they take thousands of years to make just one full circuit of the solar system. One of the best-known examples is Comet Hale-Bopp, which caused a sensa-tion in 1997—and triggered the Heaven's Gate mass suicides in San Diego. In 1950, Dutch astronomer Jan Oort suggested that the long-period comets came from a vast reservoir of ice surrounding the planets, perhaps hundreds of billions of miles beyond Neptune. This hypothetical reservoir came to be known as the Oort Cloud.[8]

Even in the 1980s, these far-flung features of the solar sys-tem were still considered hypothetical, because no one had detected any evidence of the Kuiper Belt, let alone the even more distant Oort Cloud. But in 1986, David Jewitt, a British-born astronomer who had studied comets and asteroids for years, resolved to look for Kuiper Belt objects with the same kind of dedication that drove Tombaugh's search for Pluto.

To join him in the search, Jewitt recruited Jane Luu, who had fled Saigon as a child at the end of the Vietnam War and had become a graduate student under Jewitt's wing at the Massachusetts Institute of Technology. Jewitt thought looking for the elusive trans-Neptunian objects would make a fine project for Luu's postgraduate studies. She recalled asking Jewitt why they should take on such an unglamorous search. "Because if we don't, nobody will," Jewitt told her.[9]

Jewitt and Luu began the quest much as Tombaugh did: Images of the sky were captured on photographic plates at observatories in Arizona and Chile. Then Jewitt and Luu peered at the plates using a blink comparator. It was eye-straining, mind-numbing work, which could be done for only a couple of hours at a time. Examining just one plate took eight hours.

Soon, however, the pair brought advanced technology into play. After all, these were the 1980s, not the 1930s. Arrays of integrated circuitry, known as charge-coupled devices, or CCDs, were taking the place of photographic plates for recording digital images of the sky. Those digital pictures could be run through computer processing to smooth out the images and then line them up for a quick visual scan.

The two continued their search, year after year, with better CCD and computer technology at their disposal for each new campaign. In 1988, their base of operations shifted to the University of Hawaii's Institute for Astronomy, where they had access to an 88-inch (2.2-meter) telescope atop the dormant Mauna Kea volcano, one of the world's best vantage points for scanning the heavens. Eventually, that telescope was equipped with digital detectors capable of spotting objects thousands of times fainter than Pluto.

On the night of August 30, 1992, around midnight, Jewitt and Luu finally spotted their first Kuiper Belt object. It was a faint spot that moved ever so slightly when two computer images of that part of the sky were blinked. For the rest of the night, they kept watching that spot—and the next morning the ecstatic astronomers alerted Marsden to the news. They held off on making a public announcement for a couple of weeks, however, so that they could double-check the observations. Further sightings confirmed that the faint object was far beyond Pluto. And so, in mid-September, Marsden issued the IAU circular announcing the discovery and designating the object 1992 QB$_1$.[10]

Once Jewitt and Luu figured out where and how to look with their powerful new tools, the search became easier. They found their second Kuiper Belt object six months after the first. Six months after that, they discovered two more. Other astronomers started sighting Kuiper Belt objects as well.

A whole new frontier was coming into view, thanks to improved telescopes, smarter software, and astronomers hungry for discovery. "Discovering the Kuiper Belt is like waking up one morning and finding that your house is 10 times as big as you had thought it was," Jewitt told one interviewer.[11]

Eventually, astronomers found enough Kuiper Belt objects, or KBOs, to notice patterns in their orbits. Some of the objects traced orbits in a lane ranging from 40 to 50 AU from the sun, and came to be known as classical Kuiper Belt objects. Marsden, ever the classifier, suggested that these objects be named cubewanos (pronounced like "QB-1-o's," in honor of 1992 QB$_1$, the first of its kind). Other icy worlds veered much farther out, on extremely eccentric orbits, and were called scattered-disk objects. And then there was a class of objects that, like Pluto, circled the sun in a two-to-three resonance with Neptune. Like Pluto, these ice dwarfs sometimes came closer to the sun than the giant planet, while staying a safe distance away. Jewitt suggested calling these objects plutinos—that is, "little Plutos."

These plutinos didn't trace exactly the same orbit that Pluto did. Their orbits were oriented and inclined at different angles, rather like the old pictures of electrons circling the nucleus of an atom. Nevertheless, the fact that there were other objects that followed Pluto's protected pattern—and maybe more yet to be discovered—led astronomers to the realization that Pluto wasn't such an oddball after all.

The fast-changing situation on the edge of the solar system brought the example of Ceres and the asteroid belt to mind. The latest calculations suggest that the Kuiper Belt could hold 70,000 objects wider than 60 miles (100 kilometers). That would be 300 times more than the number of similar-sized asteroids, and there might be more than a million additional ice dwarfs in the 1- to 100-kilometer range. If there were so many objects occupying the zone of the solar system through which Pluto traveled, what was so special about Pluto itself? Did it really deserve to be called a planet?

. . .

The wave of discoveries that began with Jewitt and Luu's sighting of 1992 QB$_1$ greatly expanded the suburbs of the solar system's metropolis. And it didn't take long for other astronomers to combine the power of more sensitive telescopes with more powerful computers as well.

A Ph.D. student who joined the Jewitt-Luu team in 1995, Chad Trujillo, wrote a software program that automated the blinking process for the digital imagery from the University of Hawaii's telescope. The days of sitting down at a mechanical blink comparator, as Tombaugh did in 1930, were finished. Automated blinking led to a quick upswing in the number of Kuiper Belt objects detected, not only by the Hawaii team, but by others as well.

Even Tombaugh, who was nearing his nineties, was aware of the rapid change in the celestial neighborhood he first charted more than sixty years earlier. "I'm fascinated by the relatively small 'ice balls' in the very outer part of the solar system," he wrote in a 1994 letter to the magazine *Sky & Telescope*. "I have often wondered what bodies lay out there fainter than the 17th magnitude, the limit of the plates I took at Lowell Observatory. May I suggest we call this new class of objects 'Kuiperoids'?"

Tombaugh was also aware of the looming doubts about Pluto's planetary status. Perhaps *too* aware. Ever since Marsden's talk in 1980, Tombaugh had worried about what would happen to his discovery, the brightest object in the solar system beyond Neptune. "Let's simply retain Pluto as the ninth major planet," he pleaded.[12]

In his latter years, the farm boy who became an astronomer suffered from congestive heart failure. "This controversy did not help his condition any," his daughter Annette said.

Finally, on January 17, 1997, at the age of ninety, Tombaugh's heart gave out. His obituary in the *New York Times* hailed him as the discoverer of the "ninth planet"—and said nary a word about the controversy over that title.[13]

As Pluto began its long swing away from the sun and back into the colder reaches of the Kuiper Belt, the planethood debate kept warming up. Once again, it was Marsden who brought the issue to a head: He never wavered from his view that the Lowell Observatory had put one over on the rest of the world, and that Pluto had an unjustly high status compared to his flock of minor planets. "Pluto has been a long-standing myth that's difficult to kill," the *Atlantic Monthly* quoted him as saying a year after Tombaugh's death.[14]

This time, Marsden made a modest proposal to the IAU: Thanks to the improvements in telescope and computer power, the pace of discovery was beginning to quicken, and soon the ten thousandth object would be added to the list of minor planets Marsden kept. How about giving Pluto, the misfit among the nine major planets, that place of prominence on Marsden's list? Or how about starting up a whole new list of "Trans-Neptunian Objects"? Pluto could be classified as No. 1 on that list, just as Ceres was No. 1 on the minor-planet list.

As far as Marsden was concerned, it would be okay to keep listing Pluto with the solar system's eight bigger planets as well, at least for the time being. That way, there'd be no demotion or disrespect. Pluto would merely enjoy dual status as a major planet as well as a minor planet. How would that sound?

The IAU mulled over Marsden's modest proposal, but once word got out to the public, the idea went over like a plutonium balloon. Newspapers editorialized: "Send Those Scientists to Pluto," read the headline in the *Peoria Journal Star*.[15] Letters were written by schoolchildren, including Elizabeth Bearss, a sixth-grader in Tampa, Florida: "I think Pluto should stay a planet," she wrote. "It kind of gives our solar system a personality. There's Earth, which everyone knows about; Mars, where we've sent robots; and then there's little old Pluto. He's cute and all the dogs love him, and he and Charon are inseparable."[16]

Astronomers were in the thick of the fray as well: About 135 of them quickly signed a petition opposing Pluto's designation as an asteroid, passed around by Mark Sykes, then at the University of Arizona's Steward Observatory. The American Astronomical Society's Division of Planetary Sciences issued a statement complaining that the IAU's actions would be viewed as a "reclassification" of Pluto. "We feel that there is little scientific or historical justification for such an action," the statement read.[17]

"I think Pluto's being impeached," said David Levy, the comet discoverer and Tombaugh biographer. "Pluto hasn't done anything to deserve this."[18]

Some scientists were on Marsden's side: "For at least 20 years, it's been obvious that Pluto doesn't fit," said Michael A'Hearn, a University of Maryland astronomer who was leading consideration of the policy change at the IAU's division for planetary systems sciences.[19]

Other astronomers didn't have a strong feeling one way or the other, but just felt embarrassed that so much attention was being drawn to what seemed to be a long-settled issue. "We all grew up knowing Pluto as a planet. Why upset the solar system cart at this time?" Adler Planetarium astronomer Phyllis Pitluga asked.[20]

Marsden felt stung by the whole affair. "Maybe I've been too democratic about it," he told one reporter. "Maybe I should have made the decision, and that's that."[21]

But it was too late for that. Faced with the public outcry, the IAU issued a statement denying that anyone was thinking about changing Pluto's status as the solar system's ninth planet, and announcing that Pluto would not be given a minor-planet number as Marsden wished. General secretary Johannes Andersen noted that the IAU's decisions and recommendations didn't have the force of international law, but gained acceptance only if they were "rational and effective when put into practice."

"It is therefore the policy of the IAU that its recommendations should rest on well-established scientific facts and be backed by a broad consensus in the community concerned," Andersen declared. "A decision on the status of Pluto that did not conform to this policy would have been ineffective and therefore meaningless."[22]

Plutophiles rejoiced at the news, and some of the loudest rejoicing was heard at St. Anthony's School in Streator, Illinois, the town where Tombaugh was born. Congressman Jerry Weller came by the school to praise the students' letter-writing campaign, delivering Certificates of Special Congressional Recognition. "Thanks to St. Anthony's and the student body here, you helped save the planet Pluto," he told the 189 students and their teachers.[23]

In reality, the planet wasn't in need of saving. Pluto wouldn't have gone poof if it were given a minor-planet number as well as a major-planet name. The highly public controversy did raise its profile, for a while, but the IAU's support of the status quo simply postponed the battle to come. And, unfortunately, it *was* shaping up as a battle. The controversy polarized the scientific discussion over the nature of the solar system's diverse neighborhoods, just at a time when our ability to see and study the far suburbs was dramatically widening.

Some astronomers thought the correct course was to lower Pluto's public profile. For example, at the American Museum of Natural History's Rose Center for Earth and Space in New York, the Hayden Planetarium was rebuilt in the late 1990s as an eighty-seven-foot-wide sphere that could be seen as representing the sun. Jupiter, Saturn, and other planets are displayed next to the giant ball, in sizes that reflect their scale with respect to the sun. Earth, for instance, is a ten-inch-wide sphere—about the size of a basketball. Mercury is a bit less than four inches wide—the size of a softball.

Pluto would be a couple of inches smaller, about the size of a handball. But you won't find a ball-sized Pluto mounted in the Rose Center's Hall of the Universe. Instead, a plaque labeled "Where's Pluto?" stands along the museum's walkway. The plaque explains that a ninth sphere wasn't added to the lineup simply because the display was meant to highlight the solar system's bigger classes of planets—the gas giants and the terrestrial planets. (Still, it might not hurt to leave a hand-ball lying around, just in case anyone asks).

Not even the plaque was there when the remodeled center opened in 2000. It was added only after news reports about the omission resulted in a flood of protest letters that filled up the in-box of the planetarium's director, Neil deGrasse Tyson.

One letter, from seven-year-old Will Galmot, started out with the salutation "Dear Natural History Museum" and enclosed a hand-colored picture of a blue disk in space. "You are missing planet Pluto," the letter read. "Please make a model of it. This is what it looks like. It is a planet. Love, Will Galmot."[24]

Tyson often joked about receiving stacks of "hate mail from third-graders" for leaving Pluto out of the display.[25] But even he looked forward to the day when robotic explorers could snap the first up-close pictures of Pluto and the denizens of the Kuiper Belt.

Fortunately, other astronomers were working to do just that. Even before David Jewitt and Jane Luu found their first

Kuiper Belt object, a group of planetary scientists who called themselves the "Pluto Underground" dreamt about the first space mission to Pluto. The mission went through several name changes, and it suffered a couple of near-death experiences as well. If Pluto had somehow lost its planetary status along the way, the whole effort might have gone under. But it didn't, thanks to the Pluto Underground.

7

NOT YET
EXPLORED

The way some people tell it, NASA's mission to
Pluto started with a postage stamp.

In 1991, the U.S. Postal Service chose NASA's Jet
Propulsion Laboratory in Pasadena, California, as
the site for the unveiling of a new set of twenty-nine-
cent stamps, titled "Exploring the Solar System." The ten
stamps in the set depicted all nine planets, plus Earth's
moon, and every destination was paired with a space
probe. Except one.

While nine of the stamps were adorned with a Voyager or a Mariner, a Pioneer or a Viking, a Lunar Orbiter or a Landsat, one stamp showed a simple globe—and the legend "Pluto: Not Yet Explored."

Two engineers at the Jet Propulsion Laboratory, Robert Staehle and Stacy Weinstein, saw that stamp as a challenge. They resolved to figure out how to get a spacecraft to the edge of the solar system, even if it meant shrinking the scientific payload down to a scale that didn't seem possible back then.

It's a great story—but it's only half true. Staehle and Weinstein didn't know it at the time, but a dozen astronomers who called themselves the Pluto Underground were already scheming to get NASA to send a probe to the same place. The timing was just right for the two groups to join forces and answer the challenge of a twenty-nine-cent stamp. Little did they know that it would take another fifteen years to get their idea off the ground.

The idea of sending a spacecraft past Pluto actually goes back almost fifteen years earlier than the stamp. When NASA launched the twin Voyager spacecrafts in 1977, mission managers had the option of sending Voyager 1 past a planet that at the time was known as little more than a faint sparkle in the sky.

Decision time came two years later, when Voyager 1 and its handlers faced a fork in the road: Go one way, and the spacecraft could sail past Pluto in the late 1980s. Go the other way, and it could get a good look at Saturn's rings and its cloud-covered moon, Titan. After the Saturn encounter, Voyager 1's course would take it above the solar system's plane, putting it out of range for any more planetary flybys.

Mission managers opted for Titan—a world that scientists see as an analog to primeval Earth, with hydrocarbons and complex organic molecules raining down through its thick, nitrogen-rich atmosphere. The findings from Voyager 1's 1980 flyby led NASA and the European Space Agency to start planning an even more ambitious mission to Saturn and Titan, called Cassini-Huygens, which was launched in 1997.

Pluto, on the other hand, lost out.

"Of course, at the time this decision was made, Pluto's atmosphere, its small satellites, its complex surface composition and the entire Kuiper Belt all remained undiscovered, perhaps rationalizing the Titan choice from today's perspective," said Alan Stern, the planetary scientist from the Southwest Research Institute who now leads the science team for NASA's Pluto mission.[1]

In 1989, Stern and eleven other experts on Pluto gathered in a small restaurant in Baltimore's Little Italy, after attending a seminar on their favorite scientific topic. Over pasta and wine they decided to press NASA to follow the road not taken by Voyager. The planetary scientists, who called themselves the Pluto Underground, started drumming up support for a mission devoted to the reconnaissance of what they considered the solar system's "last unexplored planet."[2]

The Pluto Underground soon had some new ammunition for the cause. In the summer of that year, Voyager 2 sent back some intriguing readings from Neptune's biggest and weirdest moon, Triton. Here was a moon that spun on a tilt, in a direction opposite from Neptune's. That suggested that Triton didn't evolve along with Neptune, but was formed somewhere else and only later was captured in a Neptunian orbit.

What was Triton most like? It had an icy, variegated surface of frozen nitrogen and methane—perhaps just like Pluto's. There were signs that geysers on Triton were spewing nitrogen and dust into its thin nitrogen-methane atmosphere—perhaps just like Pluto. The more scientists looked at Triton, the more they became convinced that Triton and Pluto were sundered cousins from the same celestial family.

On the strength of the findings from Triton, plus a flood of supportive letters from other planetary scientists, members of the Pluto Underground managed to get NASA to fund a study for a mission called Pluto 350. This concept called for a spacecraft weighing 350 kilograms, about half as big as Voyager.

Could it be done? So much downsizing would go against NASA's trend of building bigger, more expensive space robots for each succeeding mission. The two Voyager probes to the outer planets cost $865 million to build and launch. The Cassini mission to Saturn, planned as a follow-up to Voyager, cost more than $3 billion to prepare. In fact, NASA briefly considered doing a Cassini-style production, but a working group headed by Stern decided against that approach when the cost estimates ballooned beyond the $2 billion mark. The scientists favored the Pluto 350 concept, even though NASA saw it as the riskier option, and even though the trip would take twelve to sixteen years.

That was right at the time when the "Not Yet Explored" stamp made the difference. At the Jet Propulsion Laboratory, Staehle and Weinstein labored over their low-budget concept for a mission to Pluto—not knowing that the Pluto 350 team was working on the same challenge, using a different approach.

This alternative concept, known as Pluto Fast Flyby, called for building a probe that would be less than half the size of Pluto 350—140 kilograms, to be exact—and yet would still be able to answer three key questions about Pluto and Charon: What did they look like? What were they made of? What kind of atmosphere did Pluto have?

The clock was ticking on these questions. Every day, Pluto was receding farther and farther into the solar system's cold, dark depths. If the scientists waited too long, the trajectory would get trickier and the data transmissions would get dodgier. Pluto's southern winter would start setting in, leaving almost half the planet in darkness. Pluto's atmosphere might start freezing out as well. Based on computer projections, the drop-dead date would come sometime before 2020. If NASA's mission to Pluto didn't reach its destination by then, the trip probably wouldn't be worth it at any price.

So when JPL's engineers came up with a spacecraft concept in 1992 that could be launched for less money and get to its destination faster, NASA administrator Dan Goldin quickly embraced the idea. Goldin, who had been appointed the space agency's chief just months earlier, adopted the phrase "faster, cheaper, better" as his mantra for space exploration, and Pluto Fast Flyby sounded like the perfect embodiment of that philosophy. He gave the go-ahead for the development of two spacecraft that could get to Pluto in seven or eight years, at a cost of less than $500 million.

Unfortunately, Pluto Fast Flyby turned out to be the perfect embodiment of an old joke among engineers: If you're trying to make something faster, cheaper, and better, the best you can do is two out of three.

During the mission design phase, the projected weight of the spacecraft quickly exceeded the 140-kilogram target. The mission's price tag exceeded targets as well, going past the $1 billion mark. Making matters worse, NASA's planetary exploration program was thrown into disarray in 1993, when its $1 billion Mars Observer probe was lost just as it was preparing to enter Martian orbit. All these developments soured Goldin on his agency's not-so-cheap, not-so-easy mission to Pluto.

Pluto Fast Flyby went through one makeover after another. The mission was scaled back to one spacecraft instead of two. Stern worked out a deal with the Russians to launch the probe along with a Russian-built piggyback lander, but the deal crumbled when the Russians asked to be paid for the launch, which was forbidden under U.S. law. German scientists offered to step in and try to find a way around the financing snag. By then, however, Goldin had moved on to other priorities: for example, implementing a "faster, cheaper, better" approach to Mars exploration.

By 1999, NASA had spent ten years and $250 million on mission studies and hardware development, with no spacecraft to show for it. During all this time, more and more Kuiper Belt objects were being discovered. The scientific spotlight was widening to focus not just on Pluto but an entire frontier at the solar system's edge. The idea of exploring that frontier, as well as getting a close look at the "last unexplored planet," led NASA to reconsider its long-stalled Pluto plans.

This time, the mission was retooled as the "Pluto-Kuiper Express." Once again, scientific teams worked up proposals for a spacecraft that would send fresh observations of Jupiter as it flew by, then press on for an encounter with Pluto and Charon—and *then* keep sending back observations as it plunged deeper into the Kuiper Belt. But once again, the projected price tag spiraled well past the $1 billion mark.

For some at NASA Headquarters, that cost escalation was the last straw. The Pluto-Kuiper Express was abruptly eliminated from NASA's mission list in the autumn of 2000. The agency's associate administrator for space science, Ed Weiler, declared that the mission to Pluto was "over, canceled, dead."[3]

But was it really? Weiler didn't reckon on the persistence of planetary scientists—or the general public, for that matter. NASA's decision drew an outcry from experts who saw the exploration of Pluto and the Kuiper Belt as a top priority for future unmanned space missions. The Planetary Society, a space advocacy group cofounded by the late astronomer Carl

Sagan, collected more than ten thousand letters of protest and delivered them to lawmakers on Capitol Hill. That sparked a wave of questions from editorial writers and politicians.

Pluto was named by a child and associated with a children's cartoon character, so it was fitting that young people came to the tiny world's defense this time as well. A high school senior from rural Pennsylvania, Ted Nichols, created a "Save the Pluto Mission" Web site, collecting thousands of responses from around the world in a matter of days. "Don't break our kids' dream!" one posting from Osaka urged.[4]

After a few weeks of this, NASA gave Pluto another chance. Plans for a mission could go forward after all—provided that those plans posed no out-of-the-ordinary risk of failure, addressed the key goals for Pluto exploration, and cost no more than $500 million. Recalling Weiler's earlier announcement that the mission was dead, Stern quipped, "We are the undead."[5]

The Pluto project had indeed repeatedly risen from the dead, but the clock was still ticking. If the mission's backers wanted to have any chance of getting the probe to its destination no later than 2020, they had to act quickly. "Even with our best technology, it takes about 10 years to get there," Stern explained. "You can take longer, but you're not going to do it in much less time. . . . When you count the time to design it, build it and launch it, you're talking 15 years or more."

Following up on a suggestion from planetary scientists, NASA opened up a competition for the mission. Within three months, five groups of scientists and engineers submitted thick volumes setting forth their mission plans. Less than three months after that, NASA picked two of the groups to draw up more detailed proposals. For a while, it looked as if the space agency would back out of funding the Pluto mission even before a winner was named. But the mission still would not die: Congress ordered NASA to go ahead with the competition, and approved the money for Pluto over the space agency's objections.[6]

In November 2001, NASA gave the go-ahead to a mission proposal called New Horizons. Stern, a hard-driving researcher who had been in line during the mid-1990s to fly into space as a scientist-astronaut, was named the project's principal investigator. It looked as if Stern had finally succeeded in his quest to get a spacecraft launched to Pluto, twelve years after he and the rest of the Pluto Underground first gathered in a Baltimore restaurant.

NASA threw up one last hurdle in the mission's path, however. When its budget proposal for the 2003 fiscal year was released, there was no funding included for New Horizons. For the umpteenth time, Stern and his colleagues had to argue their case with Congress and NASA.

This time, the argument was cut short by a clear-cut scientific verdict—almost as if a god had descended to the stage in a Greek drama.

Every ten years, the influential National Research Council issues a "Decadal Survey" to serve as a guide for NASA on the top priorities for solar system exploration. When the new Solar System Exploration Decadal Survey came out in mid-2002, sending a probe to Pluto and beyond was the highest-ranked priority for a new mission—due in part to the Kuiper Belt discoveries that had piled up over the previous decade.

The council said the Pluto mission's findings could lead to a "new paradigm for the origin and evolution" of the solar system's little-known far frontier.[7]

That survey shattered any opposition to New Horizons, Stern said. The so-called Pluto War was finally over.[8] "Both Congress and the administration actually said, 'Oh well, if the National Academy says this is at the top—not near the top, not in the top half, not in the top quartile, but at the top—then maybe we were wrong. We need to do this,'" Stern recalled. "Congress supported it unflinchingly from that point on. So did the Bush administration, by the way. Up until that time, they did not."

Alan Stern, principal investigator for NASA's New Horizons mission, and Patsy Tombaugh, Clyde Tombaugh's widow, attend ceremonies marking New Horizons' launch.

At last, the team behind New Horizons could set to work on the spaceship that would go to Pluto. Stern's home institution, the Southwest Research Institute, was one of the team's principal partners. The other partner was Johns Hopkins University's Applied Physics Laboratory (APL), a space operation engaged in a friendly rivalry with the Caltech-managed Jet Propulsion Laboratory. APL would build the spacecraft and, once the probe was launched from Cape Canaveral, would manage the mission on NASA's behalf from its Maryland headquarters.

The 478-kilogram (1,054-pound) New Horizons spacecraft turned out to be bigger than the Pluto 350 concept that the Pluto Underground first proposed back in 1989, but still just a little more than half the size of a Voyager probe. The spacecraft generates electricity using radioactive plutonium, which is currently the only feasible power source for spacecraft that travel far from the sun.[9]

Its scientific instruments weigh just 30 kilograms (66 pounds) and draw just 28 watts of New Horizons' electricity. There's a high-resolution telescope with a built-in CCD camera, an imager that will map the composition of Pluto and Charon, a spectrometer for studying Pluto's wispy atmosphere, a radio experimental package that doubles as a communications link, devices to measure the solar wind and its interactions with Pluto, and an interplanetary dust counter. A 2.1-meter-wide (7-foot-wide) dish antenna sends data back and forth across the ocean of space.

After all the instruments and fuel were packed aboard the spacecraft, there was still room for nine mementos. Two

U.S. flags were put on board, as well as two compact disks—one encoded with 434,738 names from a "Send Your Name to Pluto" promotion, and the other containing digital pictures of the project team. Some of Clyde Tombaugh's cremated remains were placed in a canister, inscribed with a tribute to the "discoverer of Pluto and the solar system's 'third zone.'" A 100-gram piece of the SpaceShipOne rocket plane, the first privately developed space vehicle, was attached to the spacecraft. The engineers also affixed two state-themed quarters—one for Maryland, where the spacecraft was built, the other for Florida and its launch site.

And then there was the stamp: Stern made sure one of those "Not Yet Explored" stamps from 1991 was tucked inside the spacecraft.

"Pluto may not have been explored when that stamp set came out, but we were going to conquer that," he told an interviewer. "I wanted to fly it as a sort of 'in your face' thing."[10]

After seventeen years, the Pluto Underground's dream finally took flight on January 19, 2006, when an Atlas 5 rocket blasted the New Horizons probe from its Cape Canaveral launch pad into space. Among the dignitaries invited to the launch were members of Clyde Tombaugh's family and Venetia Burney Phair, the woman who gave Pluto its name when she was eleven years old. To return the favor, the dust counter on the New Horizons probe was named "Venetia."

"I feel quite astonished, and to have an instrument named after me is an honor," she said. "I never dreamt, when I was 11, that after all these years people would still be thinking about this and even sending a probe to Pluto. It's remarkable."[11]

It was a fitting sendoff for the probe, and a fitting memorial to Venetia Burney Phair as well. On April 30, 2009, while New Horizons and its Venetia dust counter were traversing the celestial emptiness between the orbits of Saturn and Uranus, Phair passed away at the age of ninety at her home south of London.[12]

To make its appointed rendezvous with Pluto and Charon in mid-2015, New Horizons took the fastest space ride NASA ever devised. Relative to Earth, its top speed was 36,250 mph (58,338 kilometers per hour), making New Horizons more than twice as fast as the space shuttle. On the way to Pluto, the spacecraft got a big gravitational boost from its encounter with Jupiter, and sent back stunning pictures of the giant planet as a bonus.

Stern and his colleagues won another bonus as well, even before New Horizons' launch. During the preparations for the mission, they were given some time on the Hubble Space Telescope to focus on Pluto and Charon—and two mysterious points of light turned up on the resulting photos. The Hubble team determined that the spots were two previously undetected moons of Pluto, now named after two creatures of the mythological underworld, Nix and Hydra.[13] As New Horizons comes nearer to its destination, still more of Pluto's dark secrets will likely be revealed. "I think it's exciting that all the textbooks will have to be rewritten," Stern said.[14]

If so much about Pluto and the Kuiper Belt is out there still to be learned, why did it take so long to get the mission off the ground?

Stern is uniquely placed to consider that question. In addition to his role as principal investigator for New Horizons, he was appointed NASA's associate administrator for the science mission directorate in 2007, filling the very post that Weiler held in 2000 when he declared that the Pluto mission was "over, canceled, dead." Stern left the space agency just a year later, after dealing with the same kinds of budgetary limitations he faced for seventeen years on the other side of the desk.

In retrospect, Stern said any mission to the outer planets faces a hard sell because of all the years of travel required to get to its destination, and the delayed gratification that is required as a result. "Bureaucracies like to do things that can take off in the time scale of those who start them," he explained. "And a Pluto mission isn't that. . . . A mission to Mars is much easier because you launch it, and next year it's there."

More than anything else, NASA's mission to Pluto came about when it did because the pace of discovery couldn't move any faster. The Pluto Underground had to wait until the revelations about Pluto's place in the Kuiper Belt finally sank in with the scientific community and created a groundswell of support.

"If we had known in 1989 what we know today—that the most populous class of planets are Pluto-like—there wouldn't have been any argument about whether it was important to go," Stern said. "We've done all this exploration, and we haven't yet been to the most populous class of planets? That's a no-brainer! We need to do that."

But would NASA have felt the same way if Pluto lost its primacy as the "last unexplored planet"? What if the International Astronomical Union had struck Pluto off the list of major planets while New Horizons' fate was still in doubt? The Pluto Underground would have lost one of its most emotional arguments for sending a probe to the solar system's edge. And even though the scientific wonders of the Kuiper Belt would still be beckoning, the public support might have faded.

"I am convinced you're on to something," Stern said, "and if the IAU had acted prior to 2003, we would probably not be en route today."

As it is, however, New Horizons is safely speeding toward its 2015 encounter with Pluto—whether you call it a planet or not.

8

BETTING ON THE TENTH PLANET

By 2003, it seemed obvious that someone was going to find a world bigger than Pluto. Mike Brown was betting on it. Literally.

The Caltech astronomer had five bottles of good champagne—Veuve Clicquot—riding on his bet with a fellow astronomer, Sabine Airieau. If an object more massive than Pluto was found beyond Neptune's orbit by the end of 2004, Brown would win the bet. If not, Airieau would get the champagne.[1]

If anyone was in a good position to find the planetary prize, it was Brown himself. In 2002, he and his colleagues happened upon the biggest solar system object discovered since Clyde Tombaugh spotted Pluto. In 2003, they outdid themselves by finding an icy world that was even larger and farther away. But neither of those objects quite measured up to Pluto, and nothing bigger came to light as 2004 was winding to a close.

Brown was starting to think he was on the wrong side of the bet. "Given that our survey has covered almost the entire region of the Kuiper Belt, I'm willing to bet these days that nothing larger than Pluto will be found in the Kuiper Belt," he told a reporter.[2]

Little did he know then that his computer's memory banks already held the imagery that would win him the sparkling wine—and spark one of the strangest turnabouts in scientific history.

Champagne wasn't the only thing at stake in Brown's bet. He was also gambling with the first several years of a promising career.

The bet really began back in 1992, when Brown was a graduate student at Berkeley, studying the volcanoes of the Jovian moon Io. One day he was walking down the hall when a postdoctoral researcher in the office next door called him over and showed him a picture of the thing she and another researcher had just discovered. The student was Jane Luu, and the thing Brown saw was 1992 QB$_1$, the first Kuiper Belt object seen beyond Pluto.

"The day before that discovery, the idea that there were large objects out there simply hadn't occurred to most people," Brown recalled years later. "And when it came time to think about what to do next, this was obviously the place to look."[3]

That time came in 1997, after Brown received his Ph.D. and was settling into his own professorship at Caltech. Once again, he happened to be walking past the right place at the right time. During a visit to Caltech's Palomar Observatory, he noticed that a 48-inch telescope was just sitting idle—and realized that this would be the perfect instrument for a planet scan. Within the year, he began searching the night sky for far-off worlds, going further down the trail blazed by Clyde Tombaugh, David Jewitt, and Jane Luu.

The trail started out along the traditional route: For three years, Brown and his Caltech colleagues slid 14-inch-square photographic plates in and out of Palomar's Samuel Oschin Telescope, capturing deep images of swaths of sky until the plates ran out. The plates' images were digitized. Then a computer sifted through the bits, looking for the telltale motions of Kuiper Belt objects. The software flagged the most promising candidates for further inspection by human eyes.

"We found absolutely nothing, but it didn't matter," Brown said. "I knew that we had the chance to find something really big and significant out there."[4]

That's when Brown bet the champagne. And that's when he took yet another gamble. Some researchers in his position

Caltech astronomer Mike Brown led the team that found the dwarf planet Eris, which displaced Pluto as the ninth-largest object known to orbit the sun.

might have taken the time to write up the negative results they spent three years of their life on, just to salvage something from the disappointment. In contrast, Brown put aside all those results and started over.

This time, he worked out a deal with NASA's Near Earth Asteroid Tracking team to pair up Caltech's 48-inch telescope with a top-of-the-line, cryogenically cooled 50-megapixel CCD camera. To assist with the computer analysis of the imagery, he brought in Chad Trujillo—the astronomer who wrote the groundbreaking software for Luu and Jewitt's Kuiper Belt search for his Ph.D., and who was now a postdoc at Caltech.

Just three months after Brown, Trujillo, and their teammates began the new search, they recorded their first hit: an icy world that appeared to be about 550 miles wide—just a little smaller than Ceres.

Six months after that, in June 2002, they found the biggest Kuiper Belt object detected up to that time. Based on its motion and brightness, the object was estimated to be more than half Pluto's size.[5] It spent most of its orbit considerably farther from the sun than Pluto, and by custom such far-flung objects are named after creation deities. Trujillo consulted with the local Tongva tribe, who lived in the area around Caltech long before the Europeans came, and the result was that the icy world was named Quaoar, in honor of "the great force of creation" in Tongva mythology.

As the largest solar system body discovered in more than seventy years, Quaoar captured the world's attention, and reenergized the public debate over Pluto's planethood. One Australian newspaper asked the question in a headline over its story about the discovery: "Quaoar, the Newest Planet . . . Or Is It?"[6]

Brown himself ventured an answer to that question: "Quaoar definitely hurts the case for Pluto being a planet," he told reporters. "If Pluto were discovered today, no one would even consider calling it a planet, because it's clearly a Kuiper Belt object."

Of course, the debate over Pluto's place in the solar system had been simmering among astronomers long before Quaoar was found. Scientific theories and observations were converging on the view that Pluto, Quaoar, and other Kuiper Belt objects were pushed out of their original orbits during the early days of the solar system.

The idea that planets didn't have to stay within well-defined orbital lanes dated back to 1984, when astronomers Julio Fernandez and Wing-Huen Ip ran a set of computer simulations that could, in effect, fast-forward or rewind the solar system's history. They found that the current shape of the solar system could best be explained as the result of a great gravitational migration that began soon after the giant planets were formed. Other astronomers fleshed out the theory over the two decades that followed.

The simulations showed that Saturn, Uranus, and Neptune likely formed closer in to the sun than they are today but moved outward because of their gravitational interactions with a thick ring of Kuiper Belt objects. Angular momentum was transferred back and forth between the outer giants and the tiny ice worlds in a complex game of orbital billiards.

The result? As the planets swept out the inner edge of the Kuiper Belt, many of the smaller objects were diverted inward, toward Jupiter. Giant Jupiter swung those interlopers back out to the farthest edge of the solar system, and since every action has its reaction, Jupiter itself moved slightly inward, shaking up the asteroid belt.[7]

Some of the Kuiper Belt objects stayed outside Neptune's gravitational grasp, although their orbits may have been disrupted. This could explain Pluto's current, fortunate orbit. Pluto most likely started out following a far more circular orbit, closer to the solar system's main plane—but was gradually pushed into an eccentric, inclined orbit that allowed it to survive for its discovery by Clyde Tombaugh.

The computer models indicated that there should be more objects beyond Neptune that were about Pluto's size. Some of the simulations even suggested that the "ice dwarfs" could get as big as Mars.[8] It was only a matter of time before telescope power and computer power rose to the level at which such objects might be found. And that's why Brown's bet seemed like such a safe proposition.

After Quaoar's discovery, the search for mini-worlds ratcheted up another notch. Another team of researchers, led Yale's David Rabinowitz, one of Brown's longtime collaborators, developed the world's largest astronomical CCD imager, a 161-megapixel camera called QUEST. The acronym stood for "Quasar Equatorial Survey Team," and one of the principal purposes for building the camera was to look for distant quasars and other curiosities far beyond our own galaxy. But it was also an unparalleled instrument for seeking dim objects inside the solar system—and in mid-2003 it was paired up with Brown's old friend, the 48-inch Samuel Oschin Telescope at Palomar.

The search process became even more automated, to the point that machines took care of the entire process of opening the telescope dome, pointing the telescope, taking pictures of the same patch of sky at three different times, digitizing the images, and sending the data to Caltech's computers. Then a computer program looked for points of light that moved just the right amount to spark a human's interest. Every morning, Brown would review the ten to twenty candidates that were flagged from the night before.[9]

The prospects would usually turn out to be false alarms—for example, bright stars that threw off tricky glints of light—but as 2003 gave way to 2004, the team identified dozens of Kuiper Belt objects, including some that rivaled Quaoar's size.

And then there was the Flying Dutchman. That was the nickname Brown and his teammates gave to an object that was first spotted in 2003 but seemed to elude further study. It was on the very edge of visibility, and faded in and out of view like the fabled captain and his ghostly ship.

The amazing thing about the Dutchman was how slowly it moved across the sky. It moved just barely fast enough to trigger the software that checked the QUEST imagery for objects worth a second look. "I just stared at it," Brown said later. "I'd never seen anything moving that slowly, and so very far away, that it was still big enough to be seen. I didn't think it could possibly be real."[10]

The slower an object moved, the farther out it had to be. And if the Dutchman was real, it promised to be the farthest-out solar system object ever found.

To nail down the object's position and orbit, the Brown-Trujillo-Rabinowitz team checked their own backlog of imagery as well as Palomar's archives. When they finally ran the numbers, they found to their amazement that the Flying Dutchman was way beyond even the Kuiper Belt—at a distance of 88 AU, or more than twice as far as Pluto. And that was nearly as close as the object would get. In six thousand years or so, the Dutchman would be at its farthest point from the sun, about 975 AU away.[11]

Temperatures on the lonely mini-world would never rise above 400 degrees below zero Fahrenheit. Brown's team played off that chilly theme by naming the object after the Inuit goddess of the sea, Sedna, whose frozen fingers were broken off and transformed into Arctic whales, seals, and walruses.

The error bars on the estimates of Sedna's size and mass are still large, but the current best guesses range between 750 and 1,100 miles for diameter, and no more than half of Pluto's mass. The most interesting thing about Sedna isn't how big it is, although its discovery did spark another rash of "tenth planet" reports. Instead, it's the mystery surrounding Sedna's strange orbit.

Astronomers were hard put to explain how Sedna got to where it is, in the inner Oort Cloud rather than the Kuiper Belt. In their paper announcing Sedna's discovery, Brown and his colleagues suggested that it could have been scattered by a yet-to-be-detected Earth-sized planet, or perhaps by a passing star.[12] Other astronomers have suggested that Sedna was actually formed around a brown dwarf, and then captured into our own solar system when the brown dwarf passed through.[13]

Brown considered Sedna's mystery to be far more scientifically intriguing than the "bigger than Pluto" controversy, and he still does. But even after Sedna was discovered, that champagne wager was still hanging over his head—and time was running out.

Sedna made Brown and his team think twice about the software they were using to cull through the QUEST imagery. If Sedna had been a little farther out, it might have been moving so slowly that the software wouldn't have flagged it. So in mid-2004 the program was tweaked to add more sensitivity. There might be more false alarms, but the planet hunters would also be less likely to miss something big.

Sure enough, the software turned up lots more far-off objects in the old image files. A particularly bright one was found three days after Christmas, in imagery that was captured on May 6, 2004. The object was given a serial number based on the date (K40506A), as well as a catchy nickname (Santa, to mark the holiday). Santa could be as big as Sedna—but it wasn't bigger than Pluto. Close, but no champagne.

On the evening of December 31, Brown e-mailed Airieau and told her she had won the bet. He went out to buy the five bottles of champagne. After the first of the year, he went back to work as usual, rechecking the archived QUEST imagery with the revised software. On January 5, he was flipping through pictures taken on October 21, 2003. Flip, flip, flip . . . then he stopped. There, in the center of the screen, was a bright spot that moved slowly—so slowly that the old software hadn't noticed it.

An object that slow-moving had to be very far away. And a faraway object that bright had to be big. Brown clicked on a button to have the computer calculate just how far away the object was, and came up with a distance farther than Pluto, even farther than Sedna: about 97 AU. Then he ran some quick calculations to estimate how big the object was, assuming that it was as reflective as Sedna. The result gave him a jolt: It could be 4,375 miles wide. That would be wider than Pluto. Wider than Mercury![14]

"I grabbed the phone and called my wife," Brown recalled. "'I just found a planet,' I said. She was pregnant at the time, and she replied, 'That's nice, honey. Can you pick up some milk on your way home?'"[15]

Brown also fired off an e-mail to Airieau: Could he have an extension on the bet? Airieau said okay—and Brown knew he would have plenty of champagne for the celebration.

That celebration had to wait a while longer, however. Brown needed to make sure that what he was seeing was real. If it was, he wanted to learn more about this object. Did it stay completely outside the Kuiper Belt, like Sedna? How big and bright was it, really? What was it made of? Brown, Trujillo, and Rabinowitz decided to keep quiet about this blockbuster until they could find more observations, nail down more of the details, and write up what was sure to be a landmark scientific paper.

The object was given its internal serial number (K31021C) as well as a sly nickname: Xena, which was borrowed from *Xena: Warrior Princess*, a syndicated TV show with a busty sword-wielding war maiden as the lead character. Brown's team had picked out that name in advance for any tenth-planet candidate that came along—partly because the "X" hearkened back to Planet X, and partly because the name took a humorous jab at the whole "name a planet after a goddess" tradition.

While the review of QUEST imagery continued, the world-hunting team came across yet another biggie in a fresh batch of observations: an object brighter than Santa, but closer than Xena. This one was designated K50331A, and because it was found just a few days after Easter, it was nicknamed Easterbunny.

The team found out much more about what Brown called the "Kuiper Belt Triumvirate" during the early months of 2005. First of all, Xena had to be downsized: It clearly wasn't bigger than Mercury. Xena's brightness had fooled Brown momentarily, because its surface was far more reflective than Sedna's. Nevertheless, it still had to be bigger than Pluto, even if its disk was made out of a perfect mirror. It also had an orbit more eccentric and inclined than Pluto's, straying farther than 97 AU and coming closer than 38 AU. That meant

Xena didn't belong to the classical Kuiper Belt but instead had been knocked into an unconventional orbit through gravitational interactions with other celestial bodies. Thus Xena is most often classified as a "scattered-disk object."

Santa was a little more conventional when it came to its orbit, but less conventional in its shape. Its variations in brightness suggested that it was rapidly spinning, which would probably give the spheroid a football-like shape. Here was one world that was as fat around its middle as the jolly old elf it was named after. What's more, Santa had a tiny moon, which the team nicknamed "Rudolph."

Both Santa and Easterbunny were about a third as massive as Pluto, but Easterbunny was smaller and brighter than Santa, perhaps because it had an icier surface composition. Easterbunny was so bright, in fact, that it could conceivably have been found during Clyde Tombaugh's sky survey decades earlier—if only it hadn't been lost in the glare of the Milky Way's celestial thoroughfare.

In mid-2005, Brown and his teammates laid out the schedule for telling the world about their Kuiper Belt Triumvirate. The findings about Santa would be shared during presentations at a planetary conference in September. That would set the stage for the big splash over Xena, about a month later. Easterbunny, which was still a work in progress, would come last.

To get the ball rolling, the team members wrote up their abstracts for the Santa presentations, short descriptions that were a traditional way to highlight a coming attraction for fellow researchers. The abstracts provided a few details about the object they called K40506A, just to whet scientific appetites, but they held back on publishing the coordinates to keep other researchers from rushing out and staking their own claims.

By July 20, all the arrangements were taken care of, the abstracts were published, and Brown eased back on his work schedule to spend some time with his wife and their newborn daughter, Lilah.[16]

That's exactly the time when the team's best-laid plans went astray.

Astronomers at the Andalusian Astrophysics Institute in Granada, Spain, had their own occasion to celebrate on July 25, 2005. An analysis of three images from the Sierra Nevada Observatory, archived since 2003, had just turned up the track of a bright object—so bright, in fact, that the astronomers suspected it could be the biggest Kuiper Belt object ever reported. After giving the images and their calculation a thorough review, they e-mailed a report on the object to Brian Marsden at the Minor Planet Center.

They also asked an amateur German astronomer, Reiner Stoss, to look for further images of the object so that its orbit could be defined more precisely. Stoss came through with pictures from other sky search programs, and eventually the object was spotted on images going back to 1955.

On July 28, Marsden sent out the center's traditional announcement about the discovery of a new minor planet, designated 2003 EL_{61}. For Brown, the e-mailed announce-ment came as a huge letdown: This was Santa, the football-shaped iceball that he and his colleagues were planning to unveil as the first of their Kuiper Belt Triumvirate. Brown took the news philosophically nevertheless, and e-mailed his congratulations to the Spanish team on that summery Thursday evening.

A few minutes later, Brown got another e-mail from Marsden, and this one was even more worrisome. Marsden had been hearing from other astronomers that this might be the ice dwarf that Brown's team called K40506A. Were they the same? Could the Spanish astronomers have been tipped off to the location of the mysterious find?

Brown's mind raced. He was pretty sure there were no leaks from the team members themselves, or from the astronomers who had been working with them. And the key information couldn't be gleaned from the abstracts. Or could it? Brown typed the serial number "K40506A" on a Google search page, just to see what came up. He was horrified to find that the number—plus Santa's coordinates—came up in a database list-ing from the Kitt Peak Observatory. Brown's team had recruited the observatory to help check for other sightings of Santa, as well as Easterbunny and Xena. The location data for all three objects was in the clear, if someone figured out the code.

After a sleepless night, Brown got on the phone to Marsden the very next morning and told him the whole story. One secret was out, and Brown decided he had to reveal the other secrets to the world before someone else beat him to it. He provided everything he had on the orbits of Xena and Easterbunny to Marsden's office, so that the Minor Planet Center could issue the discovery announcements. Xena was given the provisional designation 2003 UB_{313}, and Easterbunny was called 2005 FY_9.

Brown also made arrangements with NASA's Jet Propulsion Laboratory for a teleconference with reporters. And that's how the revelation that Pluto was no longer the ninth-biggest object orbiting our sun spilled out, helter-skelter, on a Friday evening—unquestionably the worst time of the week for announcing big news.

The question on everyone's mind was whether 2003 UB_{313} could be considered the solar system's tenth planet. When Quaoar and Sedna were found, Brown's take was that if he were defining planets, he would reserve that term for celestial bodies that were much more massive than anything else orbiting at roughly the same distance from the sun. By that definition, he said, none of the objects beyond Neptune—including Pluto—would be planets.

But when the subject was Xena, Brown took a slightly different view: "Pluto has been a planet for so long that the world is comfortable with that," he told reporters. "It seems to me a logical extension that anything bigger than Pluto and farther out is a planet."[17]

The debate wasn't purely philosophical. If Xena was considered to be like every other object that had been found beyond Neptune over the past twelve years, the discoverers would submit their suggested name to the International Astronomical

Union's Committee on Small Body Nomenclature. That committee clears the names of asteroids, comets, and anything else that is not a planet or a moon. But if Xena was a planet, the name would be reviewed instead by the IAU Working Group for Planetary System Nomenclature.

The IAU had a long list of rules for naming various objects and features. Craters on Venus, for example, should be named after famous women if they are wider than 20 kilometers (12 miles), but given common female first names if they are smaller than that. The eight planets other than Earth were named after Roman or Greek deities, but trans-Neptunian objects were named after creation deities from outside Greek or Roman lore.

The one rule the IAU did not have was: How do you decide whether a particular object is a planet or not? Astronomers had been mulling over that question ever since Marsden's assault on Pluto in 1999, and the question took on a bit more urgency when Sedna was discovered. The IAU appointed a nineteen-member panel—including Marsden and Stern as well as other experts on planets, comets, and asteroids—but they were basically deadlocked.

As long as no new object was bigger than Pluto, an answer to the question could be put off. But now, with Xena hanging over their heads and the glare of media attention shining in their faces, the IAU's top officials decided that something had to be done, and fast. The world organization's triennial assembly was coming up in less than a year, and they created a fresh panel of experts to come up with a definition by that time.

This panel would be smaller, seven rather than nineteen, so that there'd be less chance of breaking down into squabbling factions. This panel would be chaired by an eminent historian of science and include a best-selling science writer as well, so that their proposal would benefit from the historical and cultural dimensions. They would do their work in secret, so that they'd remain free from the media's meddling as well as researchers' rivalries.

This panel would come up with a way to resolve the nagging question of the past few years—well, actually, the past few decades—without discord. Or so the members of the IAU's Executive Committee hoped.

How wrong they were.

9

THE BATTLE
OF PRAGUE

I
f there's still someone out there who thinks science
and politics never mix, the story behind the Battle
of Prague should change your mind.

Some have cast the debate that took place in the
Czech capital during the summer of 2006 as a battle
against American scientists who wanted to keep the only
planet discovered by an American on an unreasonably
high pedestal. On the other side of the argument, there
are those who suspect that the rest of the world wanted

to see Pluto demoted to punish America for its unpopular foreign policy.[1]

But we're not talking about that kind of politics. We're not even talking about a battle between the fans and foes of Pluto per se. Instead of thinking in terms of Republicans versus Democrats, or Plutophiles versus Plutoclasts, you have to think in terms of planetary conservatives versus liberals—or, more accurately, dynamicists versus geophysicists. The skirmishes over the definition of planethood that took place in Prague weren't so much about poor little Pluto, but about two different ways of seeing the solar system.

One way focuses on the dynamics of a planetary system: How are things moving around, and how do those things affect one another? If a celestial body doesn't have much of a gravitational effect on other bodies, that object is hard to detect and hard to track. If lots of celestial bodies are in similar orbits, they all tend to blur together.

Pluto may be the solar system's brightest object beyond Neptune, as seen from Earth.[2] It may account for as much as 7 percent of the entire mass of the Kuiper Belt, a ring-shaped region that covers more real estate than the space inside Neptune's orbit.[3] But because there are lots of other objects in the Kuiper Belt, dynamicists see a crowded celestial neighborhood in which Pluto doesn't stand out.

Much of what astronomers have learned about the solar system since William Herschel's day has come to light because of dynamical analysis. This is how Le Verrier and Couch found Neptune. It is how Clyde Tombaugh could figure out how far away Pluto was, even though he saw it as a mere speck of light. And seventy-five years later, it is how Mike Brown identified Xena, the dynamical blip that was farther away and bigger than Pluto. So you can't really sell the dynamicists short.

Another way of looking at a celestial body would be to look *at* it rather than *around* it. What's it made of? What kinds of geological processes are at work? Does it have a crust and a core? Is there an atmosphere, and weather? Are there volcanoes, and if so, what are they spewing out? Water? Sulfur? Methane?

Such a world doesn't have to be a planet to be of interest. In fact, some of the most interesting worlds nowadays aren't planets, but moons. The Saturnian moon Enceladus is just 300 miles wide, far smaller than Pluto's diameter of 1,430 miles, but it boasts geysers that could conceivably be spewing life-laden water.

This is the province of the planetary scientists—a breed of astronomers who focus on the way a world is put together. As a rule of thumb, if it's big enough to crush itself into a round shape due to self-gravity, it's big enough to be a planet. If it's not big enough to get round, it's a failed planet, taking on the potato or peanut shape normally associated with asteroids or comets. "These objects that we call planets have shaped themselves into spheres," said Alan Stern, the planetary scientist who worked for seventeen years to get a probe sent to Pluto.

The significance of the shape isn't merely that a round object makes for a pretty, planetlike picture. Rather, the important thing is that such a degree of self-gravity makes it

An image from the Hubble Space Telescope shows Pluto with its largest moon, Charon, just below and to the right of the dwarf planet. Two smaller moons, Nix and Hydra, are visible to the right.

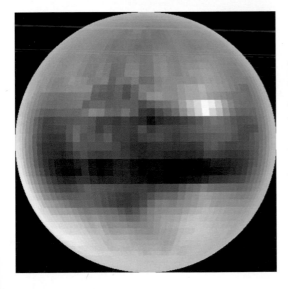

A Southwest Research Institute team led by Eliot Young constructed a color map of Pluto's surface, based on Earth-based observations of eclipses of Pluto by its moon Charon from 1985 to 1990. This view shows the hemisphere that permanently faces Charon. The red-brown color may represent hydrocarbons mixed with the surface frost.

An artist's conception shows NASA's New Horizons spacecraft flying over Pluto with its moon Charon and a distant sun in the background.

The Hubble Space Telescope captured this view of Eris, the dwarf planet that led to Pluto's reclassification, and its tiny moon Dysnomia.

An artist's conception shows Makemake, one of the dwarf planets beyond Neptune.

Solar system objects are spread over an incredibly wide volume. Scientists measure the distances in astronomical units (AU), with 1 AU equal to Earth's distance from the sun (93 million miles). The inner solar system and main asteroid belt, at far left, are contained in a disk spreading out about 4 AU from the sun. The outer solar system and Kuiper Belt, shown in the middle section, extend the disk out to 50 AU. But those distances are dwarfed by the Oort Cloud of comets and icy worlds, which is thought to form a diffuse shell with a radius of 50,000 to 100,000 AU.

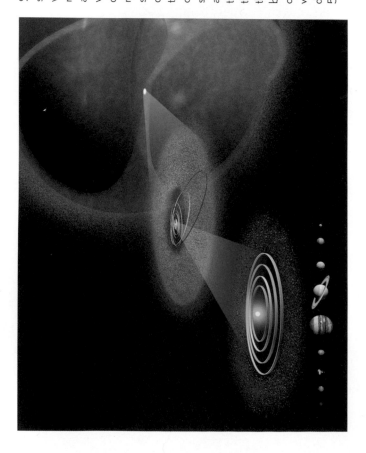

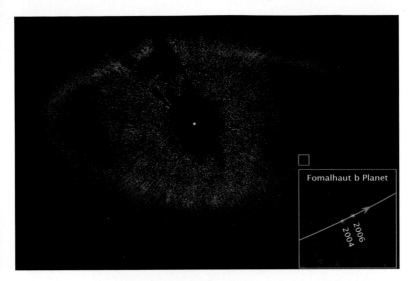

The Hubble Space Telescope spotted what appears to be a planet (highlighted in the white boxes) known as Fomalhaut b, skirting an icy ring around the star Fomalhaut.

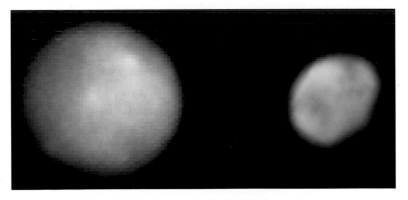

Hubble images show the dwarf planet Ceres (at left) and the asteroid Vesta (at right), the two objects targeted for study by NASA's Dawn spacecraft.

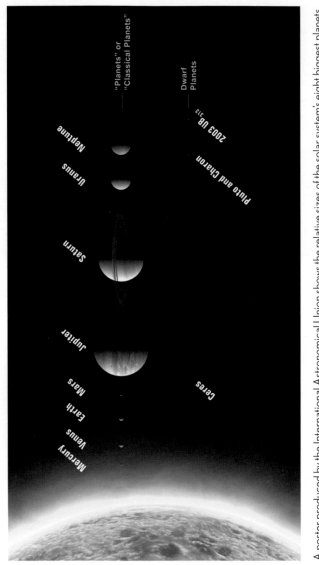

A poster produced by the International Astronomical Union shows the relative sizes of the solar system's eight biggest planets and several dwarf planets. The smallest dwarf, Ceres, looks like a speck in comparison with Jupiter.

This graphic compares the sizes of five dwarf planets and their moons with the size of Earth and its moon. The top row shows, from left, Eris and its moon, Dysnomia; Pluto and its three moons, Charon, Nix, and Hydra; and Makemake. The bottom row shows Haumea and its two moons, Namaka and Hi'iaka; Ceres; and Earth's moon. A small slice of Earth's disk is visible at the bottom of the picture.

Neptune's moon Triton, shown here in a picture from the Voyager 2 probe, is thought to be similar to Pluto in composition and may in fact have been an object captured in orbit from the same region where Pluto was formed. The surface is marked with patches of nitrogen ice and dark streaks that may have been left behind by geyser-like eruptions.

possible for a planet to have a layered composition, an active geology, perhaps even volcanic activity beneath the surface, or an atmosphere above. "It's about the physics," Stern said.

Stern likes to talk of a *Star Trek* test for planethood: "The Starship *Enterprise* shows up at a given body, they turn on the cameras on the bridge and they see it. Captain Kirk and Spock could look at it and they could say, 'That's a star, that's a planet, that's a comet.' They could tell the difference."

Roundness would provide an instant way for Mr. Spock to tell. In contrast, Stern said, having to determine whether the round thing was one object among others at the same orbital distance would force Spock to put Kirk's question on hold: "We have to make a complete census of the solar system, feed that into a computer, and do numerical integrations to determine which objects have cleared their zone."

For dynamicists, roundness just doesn't cut it. If Kirk and Spock are looking at a point of light from tens of AU away, as Clyde Tombaugh did in 1930, they might not be able to tell if the object they're looking at is round. But by closely monitoring its motion, and the motion of other bodies, they could figure out where everything fits in a planetary system—even if it takes sixty or seventy years, as in the case of Pluto and the Kuiper Belt. "We dynamicists know all about the orbits and can say what's going on," Brian Marsden said, "but the physical people can't say a damn thing."

This back-and-forth between the dynamicists and the geophysicists was what stymied the initial efforts to resolve its planet problem. Whenever the question was considered by the nineteen members of the International Astronomical Union's Working Group on the Definition of a Planet, one faction would essentially filibuster the other. "Achieving a consensus among them was about as hard as trying to herd a group of 19 feral cats into a room with several open doors and windows," said Alan Boss, an astronomer at the Carnegie Institution of Washington who was a member of the panel.[4]

In addition to the scientific differences, there was a cultural split as well, having more to do with language than physics: Should the planets of the solar system be a category so special that you can count their number on two hands, or would it be okay if the category was open-ended, with the potential of adding tens or hundreds or thousands of members?

For planetary conservatives, the idea of recognizing even thirty or fifty planets in the solar system was just too much. The liberals, however, were fine with having hundreds of planets. You could break that category down into subcategories: giants like Jupiter, terrestrials like Earth, and dwarfs like Pluto. And even if you had scores of planets, you wouldn't have to force kids to memorize them all, just as you don't force them to memorize all the world's rivers or mountains.

All these issues—the scientific as well as the cultural considerations—were dropped into the lap of a brand-new panel set up by the IAU in preparation for the Battle of Prague. This seven-member panel included five astronomers who were familiar with the issues but not counted among the leading Plutophiles or the Plutoclasts: MIT's Richard Binzel, the Université Denis Diderot's André Brahic, Junichi Watanabe from the National Astronomical Observatory of Japan, Iwan Williams from Queen Mary University of London, and the IAU's president-elect, Catherine Cesarsky. Another member was science writer Dava Sobel, the author of *Longitude*, *Galileo's Daughter*, and *The Planets*. The chairman was Owen Gingerich, an astronomer and historian who worked alongside Brian Marsden at the Harvard-Smithsonian Center for Astrophysics.

In April 2006, the committee was told to come up with a definition of planethood in time for the IAU's triennial general assembly that August, and to keep its deliberations secret, to avoid the kind of sniping that had stymied past efforts.

Gingerich tried to avoid dwelling on the particulars of Pluto's case. "We never asked who wanted Pluto in or out," he said. But the ground rules favored an approach that would lean more toward the geophysicists than the dynamicists. "We wanted to avoid arbitrary cutoffs simply based on distances, periods, magnitudes, or neighboring objects," he said.

After flurries of e-mails, the panel met in person to hash out their decision in June at the Paris Observatory, where Le Verrier had once worked to calculate Neptune's orbit. According to Gingerich, it didn't start out smoothly. "On the second morning several members admitted that they had not slept well, worrying that we would not be able to reach a consensus," he reported. "But by the end of a long day, the miracle had happened: we had reached a unanimous agreement."[5]

The resulting definition emphasized Stern's roundness requirement, but also distinguished between the solar system's "classical planets"—that is, the planets identified before 1900—and the "plutons" in the Kuiper Belt. Any world that orbited the sun and had a roundish shape due to its self-gravity, a state known as hydrostatic equilibrium, would fit under the definition of a planet.

But what if the planet's shape couldn't be seen in detail? In that case, there was a rule of thumb based on estimated diameter and mass: Objects at least 800 kilometers wide with masses of at least 5×10^{20} kilograms, or about 4 percent of Pluto's mass, would be brought into the planet fold, with borderline cases decided as further observations became available. That would put Pluto as well as Xena in the pigeonhole for planets, along with the eight bigger planets and smaller Ceres, the rocky world that was hailed as a planet in 1801 but reclassified as an asteroid decades later.

And what about Charon? Pluto's moon is nearly half as big as Pluto itself, and so, unlike every other planet, the two worlds actually orbit a common center of gravity in space, like two stars in a binary system. Some astronomers thought that would qualify Pluto and Charon as a binary-planet system, and that's what the earlier IAU panel on planethood had suggested in a footnote to their report.

"That footnote in the previous committee's report got stuck in without my quite realizing it," Gingerich said. It was one of several twists in the deliberations that he would come to regret.

Another twist had to do with the hush-hush nature of the panel's work. The IAU's Executive Committee insisted that the resolution be kept secret until the Prague meeting began. "It worked out that keeping it secret, in effect, backfired," Gingerich said. Word that Pluto would stay in the planetary fold leaked out a few days before the Prague meeting—and although the members of the panel thought their proposal would be widely accepted, others had grave doubts.

Boss recalled the tempests he and his colleagues had weathered during past discussions of the planethood issue. In an interview with the journal *Nature*, he predicted that a definition based on roundness would be met with "a long line of people waiting for the microphone to denounce it." And he was right.

More than twenty-four hundred astronomers converged on Prague for the XXVIth IAU General Assembly, which opened on August 14 and was due to last until August 25. Hundreds of panels and poster sessions were on the schedule, but none drew as much public interest as the deliberations over the definition of a planet.

On the third day of the gathering, the IAU finally published the resolution for all to see. The organizers put together a package of scientific documentation, including an explanation of the process that led to the definition, but it quickly became clear that not enough attention had been paid to the political spin.

Gingerich said his committee was "blindsided" when a press spokesman asked how many planets should be included in the proposed list. The number added up to twelve, including Pluto, Xena, Ceres, and Pluto's now problematic moon, Charon. "As soon as they started the planet-counting, that's when it all fell apart," Gingerich said.

Daniel Fischer, a German science writer who was working for the IAU's conference newspaper, was immediately struck by how poorly the organizers were stating their case. He recalls asking Gingerich to point out Ceres on a poster displaying the new twelve-planet lineup. "He looked and looked, and couldn't [find it]," Fischer said. "It was really an embarrassing moment when he couldn't find his own planet because there was dust on the poster."

Further embarrassments cropped up: The generic name that the panel had proposed for any dwarf planet found beyond Neptune—"pluton"—turned out to be already taken: *Pluton* was the name for Pluto in French, and it was also a geological term for a type of igneous rock. Also, the idea that Pluto and Charon could both be planets threw many astronomers for a loop. By that standard, our own moon, which has been slowly but steadily moving away from Earth over the course of millions of years, might someday have to be promoted to double-planethood.[6]

But the biggest political faux pas was that the dynamicists in the IAU felt slighted. They looked at the proposal and saw no reference to the issues they held most important: how much of an effect one object had on other objects in a planetary system, and how dominant an object was in the orbit that it traced. Some of them had devoted their entire careers to tracing those orbits.

Within a couple of days after the IAU panel's resolution was released, the opposition had written up a resolution of its own, one that reserved planethood only for celestial bodies that were by far the largest objects in their local populations. In a weird and fateful twist of language, Pluto, Xena, and Ceres could be called dwarf planets—but they wouldn't be considered "real" planets.

The opposition, led by Julio Fernandez, the Uruguayan astronomer who had played a part in figuring out the dynamics of the Kuiper Belt years earlier, rounded up some supporting signatures and presented its draft at a previously scheduled meeting of the IAU's Division III members, the astronomers who were most deeply involved in planetary science. In an informal show of hands, the opposition's draft won out handily over the resolution crafted by Gingerich's group.

Meanwhile, Gingerich and the IAU leadership scrambled to fix the most glaring problems in their own resolution. They got rid of the references to plutons. They wrote in an explicit distinction between the eight "classical planets" and the smaller "dwarf planets," such as Ceres and Pluto. They made clear that this definition applied only to our solar system, and not to the growing number of extrasolar planets. And they tried to fine-tune the criterion for a double-planet system.

The revised resolution looked as if it was stitched together by Dr. Frankenstein, but it served as the starting point for an open forum set for August 22, just three days before the end of the conference. The biggest practical difference between Gingerich's language and the opposition's was linguistic: Would dwarf planets be considered a type of planet or not? The stage was set for the IAU's showdown over that very issue.

When the IAU's outgoing president, Australian astronomer Ronald Ekers, called the lunchtime session to order on August 22, he reminded his colleagues that the planethood question was too big to decide by science alone. "This is not just a scientific issue of what is correct," he said. "There is no correct answer to this question. The question is, what is a sensible compromise that will work? And not just work for the professionals in the field, but work for everybody who is interested in the skies, the planets, is curious, is educating, and so on."[7]

Gingerich explained that his panel decided to "use a physical definition without an arbitrary cutoff"—that is, roundness. "It would be letting nature make the decision about what was a planet or not," he said.

When it was Junichi Watanabe's turn at the podium, he acknowledged the "extreme opinion that planets should be restricted to just eight bodies," and said he had heard the criticism that having an open-ended list of planets would be too complex for teachers to explain.

"Our solar system is already scientifically complex," he said in response. "This is a very good example to show how science is making rapid progress."

Then it was Richard Binzel's turn. He pointed out that the dividing line between planets and stars was drawn on the basis of an object's self-gravity: If the object's gravity was powerful enough to start a nuclear fusion dynamo going in its core, then it was considered a brown dwarf or a star rather than a planet. That dividing line came when the object was thirteen times as massive as Jupiter.

"Because the upper end of planets should be defined by gravity, so should the lower end," Binzel said. That's why the panel decided to draw the dividing line where an object was massive enough to crush itself into a rough sphere.

To demonstrate, Binzel used a couple of visual aids. In one hand, he held a lump of clay molded into a sphere. In the other, he held a squishy squeeze of clay. "At the lowest end, you can teach young students that this is round, it's a planet—and this is not," he said. "That's the most simple, teachable example of what we mean by the new definition of 'planet.'"

He acknowledged that there'd be a gray area where astronomers couldn't be sure how the roundness standard might apply. In that case, Division III could set the standard, perhaps based on the object's brightness.

"In summary," he said, "you can vote based on physical principle, that physics is a good way to define a planet. Or maybe you have some preconceived notion of what a planet should be. This is exactly the thing that we've been wrestling with in Division III and in our committee. Our recommendation is that you decide based on physics."

Those were fighting words for the opposition—and Italian dynamicist Andrea Milani threw the first verbal punch when Ekers opened the floor for discussion. Milani complained that the initial resolution didn't give any consideration to planetary dynamics, which was "the historically most important contribution of astronomy to modern science." He said the resolution would wrongly bring Ceres back into the planetary fold, more than a century after astronomers took it out of the solar system's top lineup. Most of all, he took offense at the way the IAU's resolution was being presented as the most scientifically sensible option.

"I would like to note that the two speakers who have spoken so far have both done the same extremely insulting gaffe," he said. "They have used the expression 'a physical definition of a planet'—by implication, suggesting that a dynamical definition is not physics!"

He said he felt he had to teach the panel "something you should know": that dynamics was indeed physics, and in fact was addressed before solid-state physics in every textbook ever written.

Milani acknowledged that the revised resolution did include more of a reference to the dynamics of the solar system but said it was too little too late. "You are perpetuating a kind of, let's say, offense to the entire dynamical astronomy community," he said.

Fernandez felt similarly hurt. He noted that he was about to become the president of the IAU's Commission 20, which focuses on the positions and motions of minor planets, comets and satellites.

"It's a pity that, occupying so seemingly high a position in the IAU, I only learned about the proposal by the Executive Committee when I arrived here, not before," he said. Then he noted that "there is an important group of people that disagree with the Executive Committee [and] have been working on an alternative proposal." He wanted assurances that the alternative would get equal treatment when it came time for a vote.

Ekers said he wanted to avoid having dueling definitions. "We would certainly rather find a compromise, rather than vote on two resolutions," he said.

As speaker after speaker came to the microphone, it was clear that a compromise would have to be struck. Some said that the resolution didn't address extrasolar planets. Some said it would be impossible to draw the line between dwarf planets and smaller objects. One complained that "inflation in the number of planets deflates, in some sense, the value of our major planets"—as if designating planets were like giving out high school grades.

Even Mike Brown, who had discovered Xena and at one point said he'd be okay with calling it a new planet, sent in an e-mail siding with the opposition—a message that was read out from the audience. "I don't know who is leading the charge, but tell them for me: I will be very happy to have my name attached to the list of supporters of the only reasonable proposal I have seen so far," Brown wrote.

The irritation in Ekers's voice rose as the objections piled up. But in the end, he was resigned to the fact that the IAU resolution was not going to fly, even in its revised form. A show of hands at the end of the meeting confirmed that impression. Ekers scheduled another meeting with the opposition, later in the day, to iron out the new wording.

"They are in control of things," Gingerich said when the session was over.[8]

Like many of the astronomers at the general assembly, Gingerich wasn't due to stay until the very end. He had airplane tickets to leave Prague on the day after the tide turned, so he missed out on the battle's final skirmishes. "Had I been there, I would have worked out a compromise," he said.

The revised definition, known as Resolution 5A, set aside any pretentions of proposing a universal definition for planethood. Instead, its scope was explicitly limited to our own solar system. Three conditions were laid out: (1) A planet had to orbit the sun; (2) it had to be big enough to crush itself into a roundish shape—that is, it had to be in a state of hydrostatic equilibrium; and (3) it had to have cleared the neighborhood around its orbit of other objects.

That third condition was the key one for the dynamicists. If a celestial body satisfied the first two conditions, but not the third, it would be considered a "dwarf-planet," complete with quotes and a hyphen. Some astronomers thought that would make clear that Pluto and the other dwarf planets were not, in fact, planets—even though the quotes and hyphen instantly fell by the wayside in popular usage.

The resolution's backers also thought the phrase "clearing the neighborhood" would be more easily understandable than talking about dominant orbital objects. But that was the part of the definition that caused the most trouble for the general public, as well as for some of the astronomers left in Prague. Could Neptune really be said to have cleared out its orbit if Pluto and the other plutinos were still buzzing back and forth? Some were even willing to argue that Pluto had done a fine job of clearing out its orbit—as evidenced by the way it survived its smash-up with the protoplanet that spawned Charon, and by the way it settled into an orbit that kept it so far away from Neptune.

To close off that argument, a footnote was added declaring that there were eight planets in the solar system, and another resolution—Resolution 6A—was drawn up stating specifically that Pluto was a "dwarf-planet" (again with the quotes and the hyphen).

Even though Gingerich had left the meeting, he made one last stab at getting some more respect for dwarf planets, by suggesting a follow-up Resolution 5B that would revive the term "classical planets" for the eight roundish things circling the sun that managed to clear out their neighborhoods. If 5B passed, that resolution could have been interpreted as recognizing that dwarf planets are really planets, too—just as our sun and other dwarf stars are really stars.

In contrast with the raucous session two days earlier, the meeting to vote on the final resolutions was surprisingly sedate. One reason for that was that the number of attendees had dwindled dramatically, to about four hundred. Another was that the clash earlier in the week had made clear who had the upper hand. As is the case with most showdowns, the outcome in Prague was determined well before the final tally was taken.

Resolution 5A, which set the solar system's planet count at eight in perpetuity, as far as the IAU was concerned, was approved overwhelmingly. To illustrate the potential impact of the follow-up Resolution 5B, one of the IAU's officers, Jocelyn Bell Burnell, brought out her own batch of visual aids: a blue balloon that represented the eight planets, a box of cereal for Ceres and the asteroid belt, and a plush Pluto toy to stand for Pluto and the Kuiper Belt. Then she placed an umbrella labeled "PLANETS" over the three props. Voting for 5B would keep the box of cereal and the Pluto toy under the planetary umbrella, while voting it down would leave them uncovered.

This vote was engineered to proceed without discussion, except for a short statement by Binzel in favor and by the Armagh Observatory's Mark Bailey against. Ninety-one voted in favor of keeping Pluto and the box of Crunchy Crisp cereal under the umbrella, but far more raised their hands and their yellow voting cards to keep Pluto out. Ekers didn't even bother to have the "no" votes counted. One of the session's biggest rounds of applause came when he ruled that Resolution 5B had failed, meaning that the IAU would no longer count Pluto as a planet.

Jocelyn Bell Burnell uses props to explain the potential impact of the vote on one of the resolutions considered at the IAU General Assembly in 2006. The balloon represents the eight biggest planets in the solar system, the stuffed doll stands for Pluto and its celestial kin, and the cereal box represents Ceres. Should all three classes of objects be covered by the umbrella provided by the word "planet"?

One loose end was tied up when the astronomers voted overwhelmingly to affirm Pluto's status as a dwarf planet and the "prototype for a class of trans-Neptunian objects." But the IAU's members couldn't agree on what to call that class: The planet definition panel's suggestion, "plutonian objects," went down to defeat by the narrow margin of 183 to 186.

Brian Marsden said he voted against the suggestion only because there was no recognition of a similar class of bodies in the main asteroid belt, which he thought could be called "cerean objects," in honor of the newly designated dwarf planet Ceres. If that change had been made, he would have voted for the resolution—and he might have persuaded one other voter to do so as well, reversing the outcome.

Nearly everyone who was there, even those on the winning side, would agree that the Battle of Prague was not the IAU's finest hour. "I think things were rather badly handled," Marsden admitted. But at least he could take comfort in the way things turned out. After twenty-six years of lobbying, he finally saw Pluto taken off the list of planets. "All things considered, we did the right thing," he said.

Astronomers vote on the resolution defining the word "planet" at the IAU's General Assembly in Prague in 2006.

His office mate, Owen Gingerich, wasn't so sure. "I realize in retrospect," he wrote later, "that the IAU should never have attempted to define the word planet. It is too culturally bound, with elastic definitions that have evolved throughout the ages. What the IAU could legitimately have done in its role of naming things was to have defined some subclasses, such as 'classical planets,' leaving the planetary door open not only for plutonians and cereans but for the exoplanets as well. These terms would be eminently teachable and would help students understand the complexity and richness of the solar system that modern science is revealing. And astronomers could have left Prague without muddle on their faces."[9]

Most of the astronomers in Prague felt they had to approve something, even if the process or the result was flawed. "It would be disastrous for astronomy if we come away from the General Assembly with nothing," the Royal Astronomical Society's Michael Rowan-Robinson told his colleagues just before the final votes. "We will be regarded as complete idiots."

So, once the votes were taken, sighs of relief could be heard in scientific circles around the world. "Now we can move on and get on with life, not argue over what is really not the major issue it's been blown up to be," said Fran Bagenal, a planetary scientist at the University of Colorado.[10]

Whether they agreed with the outcome or not, many in the scientific and educational community thought the decades-long debate over Pluto was finally over.

How wrong they were.

10

THE LIGHTER
SIDE OF PLUTO

Mercury may be a burnout case, and Mars isn't what he used to be. Venus is a hottie, but she'll make your life hell. With Saturn, it's all about the rings and the bling. Jupiter takes himself waaaay too seriously. Uranus won't stop with the off-color puns, while Neptune's jokes will leave you cold. But Pluto? Now that's one funny planet!

In the decades since Pluto was discovered in 1930, the ice dwarf was adopted as a sentimental favorite for kids and one of the most anthropomorphized bodies in the solar system, even for grown-ups.

Astrophysicist Neil deGrasse Tyson, the planetarium director who left Pluto out of the parade of planets at the American Museum of Natural History, thinks its appeal is all about the dog—the fact that generations of kids have linked the tiny world with Walt Disney's orange cartoon character. And that's certainly a factor, at least for kids like eleven-year-old Michael O'Sullivan, who was told about the International Astronomical Union's ruling during a field trip to the National Air and Space Museum.

"Seriously! Pluto is not a planet?" he asked. Then, after a moment of reflection, he added, "At least Pluto the dog doesn't have to compete with the planet anymore."

The Disney Company, which had capitalized on the new-found planet's popularity seventy-five years earlier, was philosophical about the planetary putdown. "Pluto is taking this news in stride," company spokesman Donn Walker said, "and we have no reason to believe he might bite an astronomer."[1]

But the varied reactions to the IAU's reclassification scheme indicated that the sympathy for Pluto went deeper than Disney. Pluto struck a chord as the cute little runt of the solar system's family, kicked out by the big shots of the cosmos. One editorial cartoon showed a scrawny kid pleading to be let into the treehouse for the "Solar System Club," only to be told, "Beat it, Pluto—you little ice ball!"[2]

The episode in Prague inspired comedians half a world away. "Today Pluto packed up and moved out. It said it is now going to spend more time with the family. Even sadder, it hung out around Saturn all day trying to get a job as a moon," David Letterman quipped on CBS's *The Late Show*.[3]

The perils of Pluto even provided material for one of the summer's biggest Internet sensations. "Lonelygirl15," the main character in a wildly popular faux-reality video series, snipped Pluto off a solar system mobile and related the little world's fate to her own teenage angst: "Ceres? Xena? UB_{313}? These are Pluto's new friends. Do they sound like the kids that you want to eat lunch with? No way!"[4]

Chicago stand-up comedian Ricky Marsh made it sound as if Pluto's kindred spirit wasn't Xena the Warrior Princess but rather Woody Allen: "Pluto was this little *nebbish*, never bothering anybody. Sure, it was a long distance from home, and it never called, sent a card or came for Shabbos dinner . . . but it's harmless. . . . Jews are always defending the little guy, so why should we stand by and do nothing about the inquisition of Pluto?"[5]

When it comes to Pluto's appeal, it's not all about the dog. It's all about the underdog.

There was plenty to laugh at in the way the IAU's definition turned out—particularly the definition of "dwarf planets" as nonplanets. It didn't take long for the "dwarf" label to get attached to anything that didn't quite measure up to expectations.

"How dare they say that Pluto is a Dwarf Planet?" one blurb asked on a Web site selling dwarf-planet apparel. "That's like saying Ottumwa is a Dwarf City, or like saying the Chicago Cubs are a Dwarf Baseball Team, or like saying George W. Bush is a Dwarf President . . . um, okay."[6]

Some columnists joked that the arrangement actually brought Pluto back into the planetary fold through the back door. "'Dwarf,' in addition to being politically incorrect, is only an adjective. 'Planet' is a noun, solid and palpable. Pluto, put plainly, is still a planet," Cox Newspapers' Tom Teepen wrote. "The astronomers outsmarted themselves. Which, on the evidence, may not have been all that difficult."[7]

The reactions from astronomers weren't all played for laughs: On one hand, Xena's discoverer, Mike Brown, declared that "Pluto is dead."[8] (And yet it moves.) "I'm of course disappointed that Xena will not be the tenth planet, but I definitely support the IAU in this difficult and courageous decision," Brown said. "It is scientifically the right thing to do, and is a great step forward in astronomy."[9]

On the other hand, Alan Stern—the planetary scientist who worked for a decade and a half to get a probe sent to Pluto—said the IAU's definition was "scientifically ludicrous and publicly embarrassing."

"It's going to be a laughingstock," he said. "It's going to be a mess for schoolkids. I don't think textbooks will even accept it."[10]

Clyde Tombaugh's widow, Patsy, reacted more in sadness than in anger. Since her husband's death nine years earlier, she was the one most called upon to give the perspective of Pluto's discoverer: "It kind of sounds like I just lost my job. But I understand science is not something that just sits there. It goes on. Clyde finally said before he died, 'It's there. Whatever it is, it is there.'"[11]

Patsy and other family members were the guests of honor at a protest rally at New Mexico State University in Las Cruces, where Clyde Tombaugh spent the last years of his career and his life. About fifty supporters listened to speeches and carried signs proclaiming "Size Doesn't Matter."

Meanwhile, Mark Sykes and Alan Stern started up an online petition drive, protesting the IAU planet definition as unusable and rejecting it. More than three hundred scientists and astronomers signed the petition, including some who thought Pluto shouldn't be classified as a planet but objected to the IAU's handling of the issue. The petition was removed from the Web after only a few days. "Notice of the petition began to appear on blogs, and we did not want to be swamped by signatures from the general public," Sykes said.

Lawmakers joined the fray as well, although they took their vote a little less seriously than the astronomers did. New Mexico's state House of Representatives approved a proclamation declaring Pluto to be a planet, with Patsy Tombaugh looking on from the gallery.[12] In Wisconsin, the city of Madison passed a resolution designating Pluto as "its ninth planet" while supporting "planets that take a different path, such as Ceres and Xena."[13] And in California, a tongue-in-cheek resolution condemning the "mean-spirited

International Astronomical Union" was introduced in the State Assembly.[14] (None of the measures was legally binding.)

Despite what Stern said, textbook authors at least took note of the planethood paradigm shift. In an ironic twist, Stern's own teenage son was caught up in the shift, when his science teacher marked a test answer wrong because it didn't conform to the eight-planet paradigm.

Most textbooks didn't require that much of a change, because Pluto's place had already been put in perspective thanks to the previous decade's discoveries about the Kuiper Belt. The SETI Institute's Dana Backman, who is the co-author of the textbook *Perspectives on Astronomy* as well as a researcher specializing in planetary disks, joked that he was "sort of an anti-Pluto cult leader" even before the IAU's decision. "I had always taught the material that Pluto was not a planet," Backman said. "That caused people to wake up a little bit."

In some textbooks, Pluto was wiped out entirely.[15] In others, the outcome of the IAU meeting was only half digested. For example, one set of worksheets for McDougal Littell's middle school science series asked students to add entries for five outer planets, but highlighted only the four Jovian planets (Jupiter, Saturn, Uranus, and Neptune). Pluto was still mentioned in a side note, however, as "the smallest and coldest planet in our solar system."[16]

Astronomer Thomas Arny, one of the coauthors of an introductory astronomy textbook, had to revise the chapters on the solar system twice, thanks to the IAU's pronouncements. "I honestly just don't understand what the fuss is about," he said. "There are big planets, and there are little planets. So what?"

His fellow coauthor, Stephen Schneider, just wanted to have the planethood question settled, one way or another, for the sake of the teachers. "Most of them were willing to go with the new definition—they just didn't want it to keep changing," he said.

Some teachers quickly adapted their old curriculum to reflect the new IAU view. At Jamestown High School in Virginia, Earth sciences teacher Tricia Dillon traditionally had students do up travel brochures for the solar system's planets. In the fall of 2006, the brochures carried a disclaimer: "The trip to Pluto has been canceled due to its reduced status in the solar system." Dillon used the planethood debate to teach her students about the scientific process—which, in this case, was a hard lesson to absorb. "A lot of them were sad," she said.[17]

Other teachers kept Pluto in the classroom no matter what the textbooks said. Bev Grueber, a science teacher at North Bend Elementary School in Nebraska, said Pluto is still one of the subjects handed out to her fourth-graders for oral reports on the solar system. "Everyone jumps for joy when they get Pluto," she said. "Last year, I left Pluto out of the draw and they asked where it was, so they still consider it a planet regardless of what the space scientists tell us the definition of that planet is."[18]

In an effort to raise awareness about NASA's New Horizons mission, Stern recruited a group of children called "Pluto's Pals" who were born on the same day that the spacecraft was launched. Two of those pals are twins Nora and Hana Fennell, who will be nine years old by the time New Horizons reaches Pluto. What will they be learning about Pluto by that time? Their mother, Risha Raven, couldn't predict.

"I asked their twelve-year-old sister what she thought," Raven said. "She said she's been told that Pluto was not a planet, and she really doesn't understand why."

The authors of children's books wrestled with the planetary paradigm shift as much as teachers did. Most of the newly written books reflected the IAU's view and explained Pluto's passage from planet to dwarf planet. But if you were inclined to add Pluto and Xena, you could read *Ten Worlds* to your child, and if you wanted to make sure Ceres was also covered, you could go with *11 Planets* instead.

A vigorous debate swirled around what to do with the classic memory aid for the names of the planets. The old phrase "My Very Excellent Mother Just Served Us Nine Pizzas" could be shortened by turning those "Nine Pizzas" into "Nachos." But it could just as well be lengthened by adding Ceres and Xena's new name, Eris, to the jingle: "My Very Exciting Magic Carpet Just Sailed Under Nine Palace Elephants."[19]

So how many do you memorize? One of the silliest arguments for going with eight planets, period, was that kids would eventually be forced to commit twenty or thirty or three hundred planetary names to memory. Do students have to know all 191 United Nations member states? If teachers can draw the line at remembering the five permanent UN Security Council members for social studies class, they can draw the line wherever it's appropriate for science class.

"Basically, it's a teachable moment for science teachers, because it shows the dynamic nature of science," said Gerry Wheeler, executive director of the National Science Teachers Association.[20]

If Neil deGrasse Tyson had his way, science teachers would downplay the nine-planet versus eight-planet question altogether. "The question should not be how many planets there are," he said. "There's no science in that question." The important thing is to learn about the different classes of objects in the solar system—ranging from the gas giants (Jupiter and Saturn) and the ice giants (Uranus and Neptune) to the terrestrials (Mercury, Venus, Earth, and Mars), as well as the round rocky dwarfs (Ceres) and ice dwarfs (Pluto and Xena), plus the gnarlier asteroids and comets.

The nine-versus-eight dilemma played out in spheres outside as well as inside the classroom. Musicians mostly gave up on adding Pluto to *The Planets*, Gustav Holst's famous orchestral suite. (Which was no big deal: Pluto was never part of the original music because it was written before Tombaugh made his discovery.) The National Air and Space Museum kept Pluto's plaque in its own outdoor parade of planets on Washington's National Mall, but stuck a black square over Pluto's symbol in its indoor planetary exhibit. Was the museum in mourning? It even put up an "In Memoriam: Pluto" poster, as if the 29-sextillion-pound world had wasted away to nothingness.[21]

In the toy world, some companies continued to sell mobiles and posters with nine planets, while others switched to eight. Some sold both, occasionally throwing in moons and stars as extras. *Scholastic News* artfully tried to have it both ways by offering a "Little Big Box of Planets . . . and Pluto, Too!"

When you think about it, what kid would complain about having an extra planet in the toy box? On the flip side, it's a letdown to get just eight toy planets when you were looking for nine. Blogger Jason Kottke was so disappointed to find that his fourteen-month-old son's "Solar System Ball" lacked a Pluto that he took a black marker and drew one on. "One ball at a time, people," he wrote, "that's how we win."[22]

While textbook writers, teachers, and parents absorbed what the IAU had done—and tried to figure out whether they were for or against it—astronomers tied up some of the loose ends left behind from Prague.

The top item on the to-do list was naming the dwarf planets—the very issue that sparked the Battle of Prague in the first place. After all the trouble he'd been through, Mike Brown decided the world just wasn't ready to have the biggest known dwarf planet named after a TV warrior princess. He sought a more dignified name for Xena as well as its moon, which had been detected several months after Xena's discovery. (The moon had been temporarily nicknamed Gabrielle, after Xena's sidekick on the TV show.)

Brown's choices for the official names—Eris for the dwarf planet, Dysnomia for the moon—were something of an in-joke, and a commentary on the year's turmoil. The names were taken from Greek mythology: Eris was the goddess of strife, and Dysnomia was her daughter, the spirit of lawlessness.

Eris and Dysnomia didn't conform to the IAU's custom of naming newfound objects beyond Neptune after creation deities from mythologies other than Greek and Roman. But after all, Xena had once been considered the tenth planet, and its discovery certainly brought the IAU more than its share of strife. That was a good enough excuse for the IAU working groups newly designated to sign off jointly on the names of dwarf planets—the Working Group for Planetary System Nomenclature and the Committee on Small Body Nomenclature. So, just three weeks after the IAU created the dwarf planet definition, Xena and Gabrielle were officially christened Eris and Dysnomia.

In addition to a permanent name, Eris was given a minor-planet number, 136199—signaling that, at least for the time being, the IAU's Minor Planet Center would be keeping track of dwarf planets as well as asteroids and the smaller solar system objects beyond Neptune's orbit. Pluto, too, was given a number, 134340, following through on the half-joking suggestion that Brian Marsden made twenty-six years earlier during the onetime ninth planet's fiftieth anniversary party.

Marsden felt some satisfaction that Pluto was finally being put in its place, just as he was giving up the reins of the Minor Planet Center. In fact, his last official day as the center's director was the very day that the IAU voted to put Pluto in the dwarf-planet category. "Pluto and I were retired on the same day, you might say," Marsden said.

Some astronomers thought Pluto's forgettable number might have been meant as a subtle kind of payback, coming from a person who put great stock in the significance of numbers and names. Gingerich recalled that Marsden was "swatted down" back in 1999 for suggesting a dual-status classification for Pluto. "He got his revenge by getting this extraordinarily ugly number for Pluto, which did not win him any brownie points," Gingerich said.

It definitely didn't win him brownie points from Annette Tombaugh-Sitze, the daughter of Pluto's late discoverer. "If it was the last thing he did, he was going to put an asteroid number on Pluto," she said. "And it was the last thing he did."

Another loose end was left hanging in Prague: What should the IAU call the dwarf planets beyond the orbit of Neptune? The idea was to give Pluto and the Plutophiles a consolation prize, to balance the icy world's demotion from the planetary ranks.

The first suggestion was to call them "plutons," but geologists nixed that idea. Then there was the resolution put forward in Prague to call them "plutonian objects"—a suggestion that narrowly went down to defeat. The IAU's

Executive Committee was left with the task of coming up with its own catchall term.

After more than a year of discussion, Marsden and other experts came up with the name "plutoids" to describe dwarf planets beyond Neptune's orbit, and "ceroids" for Ceres and any other dwarfs in the asteroid belt. Just as "asteroid" stood for starlike objects in William Herschel's day, the plutoid category would take in Pluto-like objects—ice dwarfs massive enough to crush themselves into a round shape. Ceroids, similarly, would be used for Ceres-like worlds.

The IAU's Executive Committee went along with the "plutoid" idea but tossed out "ceroid," declaring that scientists didn't expect to find anything other than Ceres that fit the dwarf planet category. "That is news to me," Marsden said.[23]

There were those on both sides of the planet debate who thought bringing in the new category of plutoids was just plain unnecessary. David Jewitt, the codiscoverer of the first Kuiper Belt object beyond Pluto, said he regarded the "so-called 'plutoids'" as nothing more than big Kuiper Belt objects. The only potential benefit, he said, would be if the IAU's action closed off "the equally irrelevant and politically motivated" claims for Pluto's planethood.[24]

There was little chance of that. Alan Stern, who had devoted just about as much of his life to getting a mission to Pluto as Jewitt had to studying Kuiper Belt objects, agreed that the newly coined word was irrelevant—but for a completely different reason. "It sounds like 'hemorrhoid' and it sounds like 'asteroid,' and of course these objects are planets and not asteroids," he said.[25]

Stern wondered whether this would be the last straw for those who were fed up with IAU officials and their planet definition. "They're almost needling the planetary community to go their own way," he said.

In fact, in the summer of 2008, Stern and other planetary scientists *did* go their own way, organizing a "Great Planet Debate" that finally gave the solar system's underdogs their day.

11

THE GREAT
PLANET DEBATE

For Alan Stern, the problem with Prague wasn't merely about what happened to Pluto; it was also about what happened to the scientific process.

For decades, the International Astronomical Union had worked by consensus, ruling on matters that already had been largely settled in the scientific community. The flap over Pluto and planethood was different, however, because the astronomical establishment had to deal with a basic question on a time scale that didn't fit the usual schedule for scientific consensus: How do you name something when the very naming will change the status quo?

As long as nothing bigger than Pluto had been found in the Kuiper Belt, astronomers could choose either to go with the decades-old tradition of having nine planets or trash it as unscientific. "It was that 'ninthness' of Pluto that bothered us as much as anything," Brian Marsden said.

But when Mike Brown discovered Xena, that forced the issue. Someone had to decide whether something bigger than Pluto, something that was unarguably a major planet if Pluto was, should be officially named after a grand Roman god or a lesser-known deity instead. The International Astronomical Union weaved one way in secret, and then a different way when its members pushed back during the Prague meeting. But it never considered putting off its ruling, even though some astronomers begged the organization's leadership to do so.

What bothered Stern was that there was no opportunity to look at all sides of an issue central to planetary scientists. And it bothered him even more that many scientists thought one series of votes would settle the matter.[1] "Science does not work the way the legal system works," he said. "We didn't vote on relativity or quantum mechanics. We don't vote on any scientific discovery, because it just doesn't work that way. . . . The IAU can vote that the sky is green, but that doesn't mean people will follow, because it's not."

The rift over the planet-or-not question continued long after the Battle of Prague. Eris's discoverer, Mike Brown, saw that as a bad thing. As far as he was concerned, revisiting the controversy over Pluto was like picking at a wound that should have been left to heal. "There are astronomers who want it to be a planet still, and they just keep ripping those scabs off whenever possible," he said.[2]

But Stern thought the initial operation had been botched so badly that the subject had to be left open for discussion, at least long enough for the opposing views to get a proper airing. There was even talk about setting up an alternative to the IAU. "People are asking, 'What do we need these guys for?'" Stern said. "The IAU has no special claim. They have no police force or army. They're not the Supreme Court."

Eventually, Stern decided that the best way to counter the eight-planet view was to organize a series of teach-ins for scientists and educators as well as the general public. He wanted to demonstrate that the IAU was out of touch when it came to the detailed study of the solar system.

"The fundamental issue is that not many planetary scientists even belong to the IAU," Stern said. "The vast majority of its members work on galaxies, and stars, and black holes and cosmology. The reason most of the IAU doesn't care is because it's not their issue. The people who actually understand the physics, the chemistry, the work on planets, aren't in the IAU. It's kind of like having a bunch of French professors deciding issues regarding the German language."[3]

The fact that the planet debate continued at scientific meetings over the months and years that followed—at the European Geosciences Union, the American Astronomical Society, the American Geophysical Union, and the American

Annette Tombaugh-Sitze, the daughter of Pluto discoverer Clyde Tombaugh, poses with Mark Sykes, the director of the Planetary Science Institute.

Association for the Advancement of Science, to name just a few—reinforced Stern's view that the questions swirling around what to call Pluto and its kin were far from settled.

Mark Sykes had long been allied with Stern in the debate. He was on the science team for NASA's Dawn mission, which was headed to yet another dwarf planet, Ceres. He was also the director of the Planetary Science Institute in Tucson, Arizona. In addition to his Ph.D. in planetary science, he had a law degree from the University of Arizona and was admitted to the Arizona Bar. And as if that wasn't enough, he had sung professionally onstage in more than three hundred performances as a bass-baritone in the Arizona Opera Company's chorus. All this gave Sykes an appreciation of the scientific and procedural issues of the planet debate, as well as a good sense of the drama.

In a policy article published by the journal *Science*, Sykes revived roundness as the main criterion for defining a planet: "A planet is a round object (in hydrostatic equilibrium) orbiting a star," he wrote. Why roundness? Sykes said that when an object was massive enough to crush itself into a round shape, that object also had the potential to exhibit properties of most interest to planetary scientists: volcanoes, atmospheres, eroded valleys and uplifted mountains, and even the potential for life.[4]

All these arguments received an airing in August 2008 at a "Great Planet Debate" conference, organized at Johns Hopkins University's Applied Physics Laboratory, the scientific base of operations for NASA's mission to Pluto and the Kuiper Belt. Experts representing a variety of viewpoints on the subject of planethood were invited to present and discuss their different ideas. The main event at the conference was a one-on-one debate that was broadcast live over the Internet. Sykes agreed to stand up for the dwarf planets.

Sykes's opponent for the great debate was none other than Neil deGrasse Tyson, the astronomer who kept Pluto out of the big planetary parade eight years earlier. About 150 scientists, educators, journalists, and space fans attended the Tyson-versus-Sykes talkfest, moderated by public radio host Ira Flatow.

The essence of Tyson's argument was that the Kuiper Belt had emerged as a "new swath of real estate" in the solar

Neil deGrasse Tyson, director of New York's Hayden Planetarium, received "hate mail from third-graders" when he left Pluto out of the planetary parade.

system, and that Pluto was best placed alongside the other Kuiper Belt objects, big and small, rather than alongside the giants such as Jupiter. The way Tyson told the story, Pluto would be much happier there. "It's one of the kings of the comets, rather than the pipsqueak of the planets," he said.

In contrast, Sykes said the planetary pigeonhole should make room for the eight big planets as well as the biggest of the rocky asteroids and ice dwarfs. "It's good to have a very general way of categorizing things, rather than starting out with something that serves a very narrow scientific purpose of identifying just, say, dynamical giant objects rather than looking at the problem more generally," he said. "What the IAU did wasn't to expand our perspective, but rather to narrow our perspective."

Both men appealed to history. Tyson noted that Ceres was demoted from planethood once astronomers found other, similar objects with the same orbital profile. "This is déjà vu all over again," he said. "This is just like what happened in 1801." Sykes, meanwhile, noted that at one time everything that orbited around the sun—even the smallest asteroid that could be tracked—was considered a "minor planet" by the IAU's Minor Planet Center. If you had to set a dividing line between minor and major planets, roundness would be a good standard to use, he said.

In the end, the two debaters agreed to disagree. "If you want to say planets are round . . . I don't have a problem with that. But say they're round, for hydrostatic equilibrium, and put it aside and get on with the business at hand," Tyson said.

Onstage, on that one day, the outcome looked like a stalemate. But behind the scenes, scientists were indeed getting on with the business at hand—and the scientific process was actually moving forward.

Before the IAU acted in 2006, surprisingly little research had been published on the question of how a planet should be defined. Strangely enough, one of the most thoroughgoing works on the subject was a nine-page paper prepared for the

IAU by Alan Stern and a colleague of his at the Southwest Research Institute, Hal Levison.[5]

The paper, which was written in response to the brouhaha that erupted over Pluto in 1999, laid out many of the issues that would cause so much trouble seven years later. One of the suggested definitions went for roundness, and proposed a standard based on radius and density for judging whether a celestial body was massive enough to be crushed into a round shape. The ballpark figure was a thousandth of Earth's mass, or half the mass of Pluto. That would leave Pluto in the planetary club, admit some of its kin from the Kuiper Belt, and admit Ceres as well as Pallas and Vesta from the asteroid belt.

Then Stern and Levison sketched out another standard to classify planets further, based on orbital dynamics. They noted that a line could be drawn separating planets that had cleared out their neighboring regions from planets that had not. The former category, called "überplanets," included the solar system's eight biggest worlds. The latter category, called "unterplanets," included Pluto, Ceres, and smaller round objects.

Stern said the distinction between unterplanets and überplanets was meant to be "fun and playful," making light of the mania for classification. Nevertheless, the researchers laid out a serious-sounding standard that was based on the mass of the object and its distance from the sun.

Both criteria—roundness and dynamics—ended up being combined at the IAU meeting. But instead of making the more restrictive category a subset of the less restrictive category, as Stern and Levison suggested, the IAU left the unterplanets out of the planetary picture.

How can you tell when a world is round? And how do you decide when an orbit has been cleared out? After all, there are thousands of asteroids that travel at roughly the same distance from the sun as Jupiter, lining up ahead and behind of the giant planet. And if you wanted to get technical about it, Neptune could hardly be said to have cleared out its orbit when Pluto regularly came closer to the sun. Almost no one was happy with this idea of "clearing out the neighborhood"—and so that was one phrase that cried out to be fixed.

"'Cleared' was a poor choice of terminology," said astronomer Steven Soter, a colleague of Tyson's at the American Museum of Natural History. "It confused the public, and it gave ammunition to astronomers who didn't like that definition."

Soter preferred the term "dynamical dominance"—that is, how much power a celestial object can exert on the objects around it. Even before the meeting in Prague, he wrote up a detailed paper that built on the formula laid out by Stern and Levison. By his calculation, the numbers for the eight most dominant planets came out at least a thousand times bigger than the numbers for anything else in the solar system. That power gap showed up as well in the way planets pushed around smaller objects gravitationally.[6] "You have this gap that nature is providing for us, which we can quantify both observationally and theoretically," Soter said.

By calculating the dynamical power of bodies orbiting the sun, a clear line could be drawn between the solar system's eight biggest planets and everything else—something that Stern and Levison noted years earlier.

"If it turned out that we had a thousand objects on the heavy side of the gap, they would all be planets," Soter said. "But that's not how nature did it. The solar system has room enough for only a few dynamically dominant objects—planets."

Soter's analysis showed that mass wasn't the sole determining factor for the IAU's planet definition. When it came to the dynamical effect of a celestial body, something that was far out didn't matter as much as something that was close in, even if it was as big as Mars. The analysis demonstrated that the definition depended on where a planet was and how it formed, not merely how big it was.

"If you moved Mars to Pluto's distance, it would clearly not be a planet by this definition," Soter said. "But here's the problem: You can't form Mars at that distance."

Soter and the other astronomers who were fleshing out the IAU definition thought of planets in a particular way—the way that Herschel and Le Verrier thought of them, as ruling over a region of the solar system and having an effect on the cosmic clockwork.

That wasn't the way planetary scientists such as Stern and Sykes thought of them. In fact, the claim that Mars or even Earth wouldn't fit the formula for a planet if they were farther out struck them as one of the best reasons to throw out the definition. "Any definition of a planet would be laughed out of the house unless Earth is a planet," Stern said. "Anytime you take a picture of an object, and the picture is of Earth, that has to be a planet. We live on a planet."

But how can you take a picture of a planet (or a non-planet) billions of miles away? Some astronomers saw this as the drawback to a definition based on roundness. Soter said there would be no clear dividing line between something that was round and something that was not quite round enough. "Nature is not providing us with a gap," he said.

Stern, however, said the boundary line between dwarf planets and the smaller bodies of the solar system could be drawn based on the underlying physics—that is, the measured mass and density of an object—rather than its surface appearance. "The point is that you never have to see the object or measure its roundness," he insisted. "This is a *mass* criterion that gives a *size* boundary. It's not that the object *is* round; it's that it is large enough to *be* rounded, by dint of hydrostatic equilibrium. This is very different from needing to measure roundness, or having to decide how round is 'round,' or worrying about whether an object has been hit and misshapen."

One of the astronomers who worked for Pluto's demotion in Prague, Uruguayan astronomer Gonzalo Tancredi, proposed a detailed formula for dwarf planets in a paper he cowrote with a colleague, Sofia Favre. Their analysis drew the line at a diameter of about 280 miles (450 kilometers) for icy objects and 500 miles (800 kilometers) for rocky objects. Anything smaller would be denied dwarf-planet status.

By that measure, Pluto, Ceres, and Eris were definitely dwarf planets, and another ten to sixteen objects beyond Neptune's orbit were on the candidate list.[7] Experts could quibble over the list, of course. The important thing was that there was something concrete to quibble over. "At last we have something to talk about, whereas at the time of the Prague meeting, there was not anything quantitative on these issues," said Lowell Observatory astronomer Ted Bowell, the president of the IAU's Division III.

Along with the IAU's planetary and small-body working groups, Division III's board played a lead role in fleshing out the system for naming dwarf planets—the issue that touched off the showdown over Pluto. It was this gathering of experts that came up with the name "plutoid" for all the dwarf planets that roamed beyond Neptune, as well as the naming system.

Here's where the IAU's diplomatic skills were put to the test: The names for dwarf planets had to be approved jointly by the planetary and small-body working groups, while the small-body committee alone dealt with the names for smaller asteroids and Kuiper Belt objects. This meant the IAU had to decide whether something was round without taking a picture.

The experts could go with the complicated formula suggested by the dynamicists—but that hadn't yet stood the test of time. They could go with a standard based on an estimate of mass—but for many of the Kuiper Belt objects, it was impossible to figure out the mass. Or they could base it on diameter—but here again, the size of a faraway object couldn't always be calculated to the required accuracy.

Instead, the astronomers finessed the issue by saying that the determining factor would be an object's inherent brightness, just for the purposes of naming it. If the object had an absolute magnitude of 1 or brighter, it would go through the dwarf-planet naming process. If it was dimmer, it would be considered a lesser Kuiper Belt object. All this may seem like a ridiculously complicated criterion. But it was actually a clever dodge that helped the IAU avoid—or at least defer—a fresh controversy over dwarf planethood.

The naming standard pared the IAU's list of dwarf planets down to five:

- Ceres, which was clearly round on the basis of Hubble imagery, even though it didn't meet the brightness standard.
- Eris and Pluto, which satisfied the brightness standard and looked round in the Hubble imagery.
- The two remaining members of Mike Brown's Kuiper Belt Triumvirate, which were nicknamed Santa and Easterbunny. These objects couldn't be seen well enough to get a sense of how round they were, but they satisfied the brightness standard.

All that remained, then, was to give Santa and Easterbunny their official dwarf-planet names. Brown had to cast about for more than a year to find an apt name for Easterbunny, the bright Kuiper Belt object that was found in 2005, just after Easter.

Finally, he found a reference to Makemake, the fertility god of the Polynesian people who settled on Easter Island.[8] That name made sense to Brown on more than one level, because his wife was pregnant when Easterbunny was discovered, and the astronomer said he had the "distinct memory of feeling this fertile abundance pouring out of the entire universe." So Makemake it was.[9]

Santa was a much harder case, but not because there was any difficulty in finding a name. In fact, there were too many names to choose from, due to the flap surrounding the announcement of its discovery in 2005. Spanish astronomers were the first to go public with its location. Brown, however, wondered whether they had peeked at his own observations of the curious Kuiper Belt object, which were stored in a publicly accessible online database.

Some additional server-log sleuthing indicated that the Spanish astronomers had indeed checked the database just before making their announcement—but they contended that this was done after they had discovered the object, merely with the aim of finding out whether Brown was on the same trail. Their protestations did little to ease Brown's suspicions of scientific dishonesty or fraud.

The Spanish wanted to name Santa—or, more properly, the object provisionally named 2003 EL_{61}—after an Iberian fertility goddess called Ataecina. A member of Brown's team, Yale astronomer David Rabinowitz, suggested a Hawaiian fertility goddess, Haumea. Would the IAU go with the Spanish suggestion, which came from the team initially credited with the discovery? Or would it throw out the Spanish claim as illegitimate and go with the suggestion from Brown's team?

In the end, the IAU's working groups steered a middle course. They approved Haumea as the object's name, but left Spain's Sierra Nevada Observatory as the site of discovery. The name of the discoverer, however, is left blank in Haumea's official listing.[10]

Brian Marsden said all that ambivalence was intentional. He and his colleagues tried to strike a compromise between the Spanish astronomers and Mike Brown—one of the most clever compromises devised since Solomon suggested splitting a baby in half. The Spanish said they were unhappy with the outcome, while Brown thought the dispute was resolved about as well as it could have been.[11]

"They're the discoverers, but it's his name that's being used," Marsden said. "Posterity will realize what the situation really is."

How many more dwarf planets will be named in the decades ahead? Will further evidence strengthen or weaken their claim on planethood? On these questions, too, posterity will have the final word—but the findings of the past couple of years already have focused a brighter spotlight on the solar system's far frontiers.

12

THE DAY OF THE DWARFS

Dwarf stars, like our sun, are definitely stars. And dwarf galaxies, like the Magellanic Clouds that hang around our own Milky Way, are definitely galaxies. So shouldn't a dwarf planet be a planet?

Although the International Astronomical Union ruled that dwarf planets are not "real" planets, different scientists have different ideas about how the dwarfs should be counted in our solar system. For some, it's a travesty to use the same word to describe tiny Ceres and giant Jupiter, which is two million times as massive. But that spread is no wider than the spread between the smallest galaxy and the biggest spiral galaxy.[1]

"The only difference is that the smaller objects are smaller," Alan Stern has said. "They're not fundamentally different, in the sense that a chihuahua is still a dog. A dwarf human being has all the same genetics as other humans. From my perspective, that's fine: These are dwarf planets. I coined the term, in 1991."[2]

Dwarf-planet discoverer Mike Brown says the terminology is no big deal. "The eight planets are significantly different than the many, many dwarf planets," he once wrote. "Once that distinction is made, I don't care what you call them. We could call the eight big things 'grogs' and the other 50 round things 'bloogs' and it would not matter to me."[3]

But a distinction also has to be made between the bloogs—er, the dwarf planets—and the hundreds of thousands of other objects in the asteroid belt and beyond Neptune. The dwarfs are definitely built differently from their smaller kin. The three largest objects in the asteroid belt—Ceres, Pallas, and Vesta—are more firmly packed than any other objects in that zone. The same can be said for Eris, Pluto, and Haumea in comparison with the less dense objects outside Neptune's orbit.[4] Lumping them all together—for example, by saying that Eris is just a big trans-Neptunian object or that Ceres is just a big asteroid—glosses over the differences that put the dwarf planets in a class by themselves.

Here's a mini-guide to the first five mini-worlds singled out by the IAU, based on the latest information and speculation:

Pluto

Distance from sun: 29.7 to 49.3 AU

Equatorial diameter: 1,430 miles (2,302 kilometers)

Mass: 1.3×10^{22} kilograms, 0.002 Earth mass or 17 percent of the moon's mass

Orbital period: 248 years

Although it's been called an iceball, Pluto is actually thought to consist of roughly 65 percent rock and 35 percent ice by mass. By volume, it's about half-and-half rock and ice. Planetary scientists believe the rock has settled into a central core, surrounded by a mantle of water ice. Some of the water in contact with the rock may exist as liquid, due to warmth given off by radioactive decay. The water, ice as well as liquid, could contain the chemical building blocks for life, perhaps the same types of building blocks that were put together on Earth. The possibility that life arose on Pluto or other ice dwarfs is small. But it's not quite zero.

Pluto's surface is coated with nitrogen ice, plus bright traces of frozen methane and carbon dioxide. We've known for decades that it has a thin atmosphere of nitrogen with traces of methane. Just recently, scientists have found that Pluto's atmosphere is warmer than its surface, due to a temperature inversion.[5] In fact, there seems to be an atmospheric cycle that periodically deposits fresh frost on the ground. Will the atmosphere freeze out completely when Pluto swings farther from the sun? Scientists assume so, but it's a question they're anxious to answer definitively.

They're also anxious to learn more about Charon, Pluto's biggest moon. Some still argue that Pluto and Charon make up the first known double-planet system, because they orbit a center of mass that lies between them.

Charon appears to be icier than Pluto—which would make sense if the two bodies were formed as the result of a primordial collision. The impact would have blasted the ice from Pluto's top layers into space, and computer models suggest that ice could have helped shape Charon.

Charon doesn't seem to have an atmosphere, but it does show evidence of having volcanoes or geysers that spew water ice. If that evidence is confirmed, it would lend weight to the idea that water has remained in a liquid state deep beneath Charon's surface—perhaps due in part to the presence of ammonia, which would act as natural antifreeze.

Two other tiny Plutonian moons, Nix and Hydra, were discovered in 1995. They are probably bits of ice left over from the collision that gave rise to Pluto and Charon. Pluto thus holds sway over more moons than any of the four terrestrial planets.

Eris

Distance from sun: 37.8 to 97.6 AU

Equatorial diameter: 1,500 miles (2,400 kilometers), based on Hubble Space Telescope observations

Mass: 1.67×10^{22} kilograms, or 27 percent more massive than Pluto

Orbital period: 557 years

Named after the Greek goddess of discord, Eris may have been seen in some circles as the "Pluto-killer"—but she's actually more like Pluto's slightly bigger sister.

The two dwarf planets' diameters are nearly the same, as is the basic recipe: rock in the middle, sheathed in ice, with a coating of frozen methane and nitrogen on top. Eris is 27 percent more massive—in part because of the size difference and in part because Eris is slightly rockier and less icy than Pluto.

The most significant difference is in the orbits. If you thought Pluto was an oddball, consider this: Eris is currently three times farther away from the sun than Pluto, but periodically comes closer to the sun than Pluto. Its orbit is inclined 44 degrees to the solar system's main plane, compared with Pluto's 17-degree inclination.

Since Eris's discovery in 2005, planetary scientists have seen signs that the balance of methane and nitrogen varies on the surface. For some, this suggests that Eris is experiencing some sort of seasonal weather change. Perhaps nitrogen and methane are taken up into Eris's thin atmosphere when the sun's feeble light warms one side of the surface. Then those gases are transported over to the dark side, to be deposited as frost.

Another possibility is that methane and other volatiles are welling up through ice volcanoes—perhaps along with liquid water that has been laced with ammonia, as scientists have suggested is the case for Charon.

Does Eris have weather? Or ice geysers? Or both? Further observations could tell the tale, now that planetary scientists have a better idea of what to look for.

Eris's moon, Dysnomia, could shed additional light on the dwarf planet's origins. Dysnomia may have been formed the same way Charon and our own planet's moon are thought to have come into existence—that is, as the by-product of a cosmic collision.

Haumea

Distance from sun: 34.7 to 51.5 AU
Equatorial diameter: 1,225 by 950 by 622 miles (1,960 by 1,518 by 996 kilometers)
Mass: 4.2×10^{21} kilograms, or 32 percent of Pluto's mass
Orbital period: 285 years

The dispute over naming Haumea isn't the only thing that's controversial about the ice dwarf formerly known as Santa. It doesn't look spherical at all, but more like a squat loaf of sourdough bread. So how can Haumea possibly meet the "roundness" standard for dwarf planets?

The reason is that Haumea rotates six times a day, creating a spin that forces the material outward at the equator. Although it looks squashed rather than round, Haumea is thought to be in a state of hydrostatic equilibrium because its shape is determined by self-gravity rather than the structure of the materials it's made of.

This phenomenon isn't unusual. All eight of the solar system's dominant planets have a somewhat flattened shape. Saturn's pole-to-pole diameter, for instance, is about 10 percent shorter than its diameter at the equator. For Earth, the difference is less than 1 percent. Haumea is an extreme case, with a polar axis that's about half of its maximum width at the middle.

Haumea is different from Pluto and most of its other brethren in the Kuiper Belt in that it's made almost entirely of rock, with just a thin veneer of ice right at the surface. This is what led Mike Brown's team to name the dwarf planet after a Hawaiian fertility goddess associated with earth and stone.

Haumea's rocky composition and its strange spin cycle have sparked speculation that it was blasted to bits by another Kuiper Belt object long ago. The collision set Haumea's denser, rocky stuff whirling into its current shape, like dough in a food processor.

Smaller, lighter bits of icy debris from the collision would have separated from the main mass—and at least two of those bits now orbit Haumea as moons. The moons are named Hi'iaka and Namaka, after two lesser Hawaiian deities that sprang from Haumea's body. Brown's team says several other icy bits went into their own orbits around the sun, forming a collisional family that can be traced back to Haumea through computer simulations.

Haumea and all her progeny are stuck in a particular type of orbits that are fated to become more and more eccentric over time, due to gravitational interactions with Neptune. It could well be that some of the icy bits have already become short-period comets. Brown suspects that Haumea itself could turn into a comet one day. "When it does it will probably be 10,000 times brighter than the spectacular comet

Hale-Bopp, making it something like the brightness of the full moon and easily visible in the daytime sky," he says.[6]

For all these reasons, Brown rates Haumea as his "favorite object in the solar system."

Makemake

Distance from sun: 38.5 to 53.1 AU

Equatorial diameter: Estimated at 900 miles (1,500 kilometers)

Mass: Estimated at 4×10^{21} kilograms, or roughly 30 percent of Pluto's mass

Orbital period: 309.9 years

Discoverer Mike Brown has called Makemake the "Rodney Dangerfield of the larger objects in the outer solar system" because, like the late comedian, it gets no respect. Unlike Haumea, it doesn't have moons or an intriguing shape. Unlike Pluto or Eris, it doesn't appear to have an atmosphere. But Makemake does have some qualities that make it a standout in the night sky.

The dwarf planet is covered in virtually pure hydrocarbons: frozen methane and ethane, with little sign of nitrogen. Some of the methane appears to have been altered by the sun's ultraviolet light, creating chemicals known as tholins that give Makemake a reddish hue. This suggests that Makemake lost much of the nitrogen in any atmosphere it may have had, and that the hydrocarbons fell out of the atmosphere as frost.

Because hydrocarbon frost is highly reflective, Makemake is one of the intrinsically shiniest objects in the solar system. It's currently right behind Pluto as the second-brightest object in the Kuiper Belt, as seen from Earth.[7]

If Clyde Tombaugh knew exactly where to look, he might have found Makemake back in the 1930s during his survey for Planet X. But at the time, the ice dwarf was passing against the star-filled background of the Milky Way. That would have made it devilishly hard for Tombaugh to spot.

Some scientists have speculated that finding Makemake back then, or even back in the 1950s, would have changed the character of the decades-long debate over Pluto's status. Such "what if" stories abound in the history of science. Just imagine, for instance, how the course of astronomy might have changed if Galileo had recognized Neptune as a planet back in 1612. Or imagine how much less we might have known if Tombaugh hadn't been driven to look for the mythical Planet X in 1929. Sometimes what you discover is determined by what you're looking for—which is a good argument for keeping an open mind about the breadth of the solar system spectrum.

Finding Makemake wasn't easy even for Brown's team. Its steeply inclined orbit brought the object far north of the ecliptic plane, which had been the usual hunting grounds for planetary objects. Now Brown and his colleagues realize that every inch of the night sky will have to be scoured for evidence of dwarf planets.

Ceres

Distance: 2.55 to 2.99 AU
Equatorial diameter: 606 miles (974.6 kilometers)
Mass: 9.4×10^{20} kilograms, roughly 25 to 35 percent
  of the main asteroid belt's total mass, or 7 percent of
  Pluto's mass
Orbital period: 4.6 years

Ceres is the odd one out among these first five dwarf planets:
It's the only one located in the asteroid belt rather than the
Kuiper Belt. And although a few other asteroids are biggish
and roundish, Ceres may be the only asteroid ever admitted
into the dwarf-planet club. Some have even suggested that it
stands out from the asteroid crowd so much that it may be an
interloper from the Kuiper Belt.[8]

Like Pluto, Ceres will be getting a visitor in 2015: the Dawn
spacecraft, which is due to go into orbit around the dwarf
planet after visiting Vesta, the asteroid belt's second most massive object. The findings from Dawn could well raise some
eyebrows: Ceres is thought to have a rocky core and a mantle
of water ice, covered over by a crust of clay and dust. This has
led some researchers to call it an "embryonic planet" whose
development was put on pause. "Gravitational perturbations
from Jupiter billions of years ago prevented Ceres from accret-
ing more material to become a full-fledged planet," University
of Maryland astronomer Lucy McFadden says.[9]

Under just the right conditions—for example, if Ceres's
subsurface water layer is laced with ammonia antifreeze, and

if the core is giving off radioactive heat—liquid water may still lurk in the dwarf planet's interior. But even if Ceres's water is frozen solid today, it might have stayed liquid long enough when Ceres was young to support organic chemical processes, and perhaps life.[10]

For that reason, the Dawn team has ruled out any scenarios that call for crashing the spacecraft on the dwarf planet's surface, for fear of contaminating a potential target for future biological study. "We go into a quarantine orbit at the end of the mission," the Planetary Science Institute's Mark Sykes says.

The IAU's planet definition may have pushed Pluto a little farther out of the spotlight, but at the same time, it has given Ceres a bigger role in the planetary play. No longer seen as just one rock amid thousands of other bits, Ceres has regained some of the prestige it had back in 1801 when Giuseppe Piazzi hailed it as what was then the eighth planet. Today, Ceres has a revised place in history as the first dwarf planet to be discovered, and the smallest of the lot.

More to Come

How many more dwarf planets will be added to the list? Theoretically, there could be scores, or hundreds, or thousands. Critics who say the classification scheme is too broad point to the Saturnian moon Mimas—which is only 250 miles (400 kilometers) wide, and mostly round. That is, if you don't count the big bull's-eye crater that makes it look like the Death Star from the *Star Wars* movies.

If Mimas were in its own orbit, it might be considered a dwarf planet. But what's the harm in that?

The critics complain that there might be thousands of 250-mile-wide objects in the Kuiper Belt, and that the list of planets might spin out of control. But as long as a distinction is made between the eight dominant planets and the lesser worlds, what's the harm in that?

Practically speaking, even the dwarf-planet registry will likely expand slowly. The first measure of dwarf-planetary status, based on brightness, is relatively easy to calculate. If newly discovered objects have an absolute magnitude that is dimmer than, say, Charon, they won't automatically be added to the dwarf-planet list.[11] In order to add those dimmer objects, astronomers would have to present evidence of hydrostatic equilibrium—for example, direct observations of the objects by next-generation telescopes.

Given the right kind of exposure, the myriad of planetary bodies could easily become as well-known to children as, say, the myriad of dinosaur species. There are already enough such worlds for a respectable set of trading cards. The five—or fifty, or five hundred—dwarf planets should become as much a part of the full set as the solar system's eight trump cards.

And Alan Stern has suggested that if we hang around the solar system long enough, the dwarf planets may become the only game in town. Billions of years from now, when the sun puffs itself up into a blazing red giant, the ice dwarfs would lie in what Stern calls a "Delayed Gratification

Habitable Zone." In such a scenario, Earth and the other inner planets would be turned into cinders at best, while worlds as far out as 50 AU might just offer a temporary foothold for life.

"When the sun is a red giant, the ice worlds of our solar system will melt and become ocean oases for tens to several hundreds of millions of years," Stern has said. "Our solar system will then harbor not one world with surface oceans, as it does now, but hundreds."

Other scientists aren't as optimistic about the prospects for beachfront property on Pluto. Once the sun enters its red giant phase, the entire solar system would be thrown into chaos, and it's not clear whether any region would be stable enough for long enough to sustain life. "The idea of organic-rich distant bodies getting baked by a red giant star is an intriguing one, and could provide very interesting if short-lived habitats for life," said Donald Brownlee, an astronomer from the University of Washington who has delved into the prospects for life elsewhere in the universe. "But I am glad that our sun has a good margin of time left."[12]

In the meantime, astronomers will be using increasingly powerful telescopes to explore the wide realm beyond the Kuiper Belt—the mysterious region of the solar system where Sedna was discovered in 2004. Sedna may be the first object found on an astronomical frontier where an even bigger Planet X still waits to be discovered.

13

PLANET X REDUX

It took more than seventy years to find a new world in our solar system that was bigger than Pluto, but it shouldn't take anywhere near that long to find the next "Planet X." Statistically speaking, the chances are good that somewhere on the solar system's edge, something bigger than Mercury or perhaps even Mars is lurking.

After all, if the oceans of space beyond Neptune contain a respectable number of objects around Pluto's size, there should be at least one or two that are substantially bigger. Our telescopes have yet to plumb the full depths of the Kuiper Belt—let alone the vast Oort Cloud beyond, which serves as the source of the solar system's most distant comets. Astronomers can't predict exactly what might be found, or exactly where, but most are confident that there's something big out there.

This fascination with planets yet unseen has spawned a modern-day doomsday myth dressed up in ancient lore. According to the doomsayers, a giant planet referenced in Sumerian texts, dubbed Nibiru, is making its way through the solar system and will set off a planetary catastrophe when it passes by Earth.[1] The Nibiru tale has been rolled up with yet another myth tied to the end of a Maya calendar cycle in 2012, resulting in a double dose of doomsday.

"It's disheartening, because people are really frightened," NASA astronomer David Morrison said. As senior scientist for the NASA Astrobiology Institute at Ames Research Center, Morrison has had to cope with hundreds of e-mailed questions about Nibiru, the talk about 2012, and claims of a government cover-up.

Despite Morrison's repeated assurances that there's nothing to fear, the rumblings about a malevolent Planet X continue—fed in part by the real science surrounding the search for worlds beyond Pluto. Some point to an infrared sky survey back in the 1980s that turned up anomalous readings (which were later traced to distant galaxies).[2] Others cite theoretical studies that suggest how distant worlds could divert comets toward the sun (even though the studies make clear that Earth's orbit would not be disrupted). Still others wonder if the dwarf planets are the messengers of doom foretold by the Sumerians (a worry that was sparked by all the news reports about Eris, initially nicknamed Xena or Planet X).

"Maybe we should be asking about Eris and not Nibiru," one questioner told Morrison in an e-mail. "Thank you for your time, as I am scared to death!"[3]

The reality is that scientists see no signs that any planet is coming to get us. But that's not to say all the mysteries have been resolved. As astronomers learn more about the solar system's icy frontier, they are asking deeper questions as well: How did today's planetary zones develop? Are there objects out in the Oort Cloud that send storms of comets our way? How do you explain Sedna, an icy world that takes twelve thousand years to complete just one orbit and comes nowhere near the Kuiper Belt?[4]

Today, Sedna and its kind, commonly known as distant detached objects, are the solar system's oddest oddballs. One of the giant planets—say, Neptune—could have kicked Sedna outward into a crazily eccentric orbit. But changing the orbit again so that it never comes back would require another gravitational kick from something big on Sedna's far side. Could there be planets that big yet to be discovered, hundreds or thousands of AU from the sun?

Several teams of researchers have concluded that such worlds are plausible, even though no one can yet say whether they actually exist. Astronomers in Japan, for instance, used computer modeling to determine that a world somewhere between the mass of Mars and Venus could explain all the weird orbits of the objects beyond Neptune—that is, if it were 80 AU or farther from the sun. That would be roughly twice Pluto's distance from the sun, but the mystery world could still be detected from Earth during the closest part of its orbit.[5]

"I would like the public to understand that the research on distant hypothetical planets is still active (including my own) and that several questions remain open yet," said one of the astronomers, Patryk Sofia Lykawka. "In addition, it is important to understand that such theories in planetary sciences have absolutely no relation with Nibiru, 2012 or other hoaxes that claim for the existence of 'apocalyptic' or 'mystic' celestial bodies."

Another research team has proposed the existence of an object as big as Neptune or even Jupiter, placed fifty to one hundred times farther out from the sun than Pluto.[6]

In all these cases, the suggested location of a Planet X is based not on any actual observations, but on what it would take to get the right results in the computer simulations—essentially, a twenty-first-century mathematical exercise akin to the reverse calculations made by Le Verrier and Lowell in the nineteenth and twentieth centuries.

Some astronomers are looking at an even bigger picture: Could the weird goings-on at the edge of the solar system have resulted from a close encounter between our sun and another star? At the Carnegie Institution of Washington, planetary scientist Alan Boss suggests that Uranus and Neptune could have been sculpted into their current forms by ultraviolet light coming from a hot, young star passing through the cosmic neighborhood.[7]

Such a star also could have diverted some of the Oort Cloud's icy worlds into the weird orbits we see today. But because we don't see any stellar hotties in our cosmic neighborhood today, this scenario supposes that the hot star left the scene, leaving behind the mysterious planetary re-arrangement as its calling card. Some even worry that a stellar intruder has made periodic trips through the solar system, stirring up storms of comets and causing mass extinctions on Earth. That's sparked yet another set of doomsday worries about a supposed "Death Star" that's been nicknamed Nemesis.[8]

John Matese, an astrophysicist at the University of Louisiana at Lafayette, has been looking into the Nemesis scenario for more than two decades. In collaboration with a colleague at Lafayette, Daniel Whitmire, and NASA researcher Jack Lissauer, Matese has traced the ebb and flow of cometary impacts over billions of years. He agrees with the widely accepted view that the main factor behind the ebb and flow is the solar system's movement through the galactic tide, an up-and-down bobbing motion that takes millions of years to play out. As far as he knows, there's no sign of a Nemesis. But he's not willing to rule out a giant Planet X just yet.

"If there is anything in the Oort Cloud that is a cause of this suggestive data, this companion would have to be relatively massive—something on the order of three to ten Jupiter masses, with its mean position out at a distance of 10,000 AU," he said. "But the data just isn't good enough at the moment to go beyond saying it's suggestive."

Matese said that if such a Planet X does exist, astronomers should be able to detect it once a satellite known as the Wide-Field Infrared Survey Explorer (WISE) makes its full scan of the sky. "If it doesn't discover it, then the whole discussion should be concluded," he added.

WISE is just one of several space projects that could bring about fresh breakthroughs in the search for planets, from A to Z:

- **Discovery Channel Telescope:** Lowell Observatory, the place that made Pluto famous, has partnered with the Discovery Channel to build a $40 million telescope in Arizona that will extend the search for Kuiper Belt objects, as well as extrasolar planets and near-Earth asteroids.
- **Giant Magellan Telescope:** The $700 million GMT, due to be built in Chile by 2018, will combine the power of seven 27.6-foot-wide mirrors to produce images sharper than those of the Hubble Space Telescope. The instrument should be able to see the disks of Sedna and more of its faraway kin, piecing together the evidence for or against the existence of a giant Planet X—or even a Nemesis-type star that may have passed by during the solar system's infancy.[9]
- **Large Synoptic Survey Telescope (LSST):** The $400 million LSST is expected to become fully operational in Chile in 2016. "In the first week, we will see more data from this telescope than all the telescopes in humanity

up to that point," billionaire backer Charles Simonyi says. The LSST is expected to spot up to 100,000 orbiting objects beyond Neptune, including ice dwarfs as big as Pluto out to a distance of 200 AU. Among the researchers involved in the LSST project is dwarf-planet discoverer Mike Brown.[10]

- **Panoramic Survey Telescope and Rapid Response System (Pan-STARRS):** The $100 million Pan-STARRS is an array of four telescopes being set up in Hawaii primarily to track fast-moving asteroids, some of which might threaten Earth. However, Pan-STARRS is expected to spot about 20,000 Kuiper Belt objects and should be able to find objects as small as Pluto well beyond the belt. In fact, a Planet X like Jupiter could be seen at a distance of 2,140 AU—more than fifty times farther away than Pluto. One of the leaders of the Pan-STARRS effort is the University of Hawaii's David Jewitt, who co-discovered the first Kuiper Belt object beyond Pluto back in 1992.[11]

Brown estimates that as many as 200 dwarf planets could be found in the Kuiper Belt, plus another 2,000 or so when the Oort Cloud is surveyed.[12] But if a world the size of Mercury or Mars is found, as Brown and most other astronomers expect, that planet would be a "dwarf" in name only.

The American Museum of Natural History's Steven Soter thinks such objects would require a new label. "We might well find Mars-sized or larger objects in the outer Kuiper Belt or

the more distant Oort Cloud," he said. "If so we would probably conclude that such objects were formed in the inner solar
system before the gravity of the giant planets tossed them
into the outer regions. Such bodies would not be dynamically
dominant and, to distinguish them from the regular planets,
we might call them 'scattered planets.'"

Brian Marsden, the consummate planet classifier, said
the status of far-off objects the size of Earth would depend
on the shape of their orbit. If they stayed well beyond the
Kuiper Belt, in well-behaved, cleared-out orbits, they could
be regarded as planets under the International Astronomical
Union's definition.

"If we do find something like that, I think that's much more
likely that we would consider them another set of planets,
just as Jupiter through Neptune are actually different from
Mercury through Mars," he said.

But if a distant world's orbit was so eccentric that at its closest point it dipped closer to the sun than Neptune, it would
be disqualified—at least in Marsden's book. "We would have
to say that that Earth-sized object is not a planet," he said.[13]

The hypothetical case of a dark Earth orbiting far beyond
Neptune sheds a different light on the exercise of defining a
planet on the basis of not only where it is, but where it eventually goes. For critics of the IAU definition, this is a fatal flaw.
"The one thing that dynamicists realize would topple the IAU
apple cart altogether would be to discover a trans-Neptunian
object bigger than Mars," Mark Sykes said. "And that is quite
possible!"

As if that's not mind-boggling enough, there's the question of what to call planet-sized objects that are outside the orbital influence of a star, either because they were kicked out of their planetary nest or because they formed in isolation. Some of these objects are even thought to have moons (or are they subplanets?).[14] If you're an astronomer who's a stickler about terminology, you'd deny these objects the planet label and call them sub–brown dwarfs or "planemos" (short for planetary-mass objects). But for most people, even for most astronomers, the term "rogue planet" will do just fine.

The oddball planets demonstrate how handy it is to have adjectives at the ready, for dwarf planets and dominant planets as well as scattered planets and rogue planets. "The word 'planet' by itself doesn't give us enough information to think critically about what someone is telling us," Vanderbilt University astronomer David Weintraub said. "Jovian . . . Neptune-sized . . . Earth-sized. . . . You almost have to have those adjectives in order to make the word 'planet' useful anymore."

Why is quibbling over one noun and a few adjectives so important? One answer is that the concept of planethood plays such a key role in the deepest questions we have about the universe, questions that range far beyond our own solar system. Are we alone? Could alien planets harbor life? Could they become future homes for our descendants, even if it takes millions of years for them to get there?

Scientists have detected more than three hundred planets beyond our own solar system, and the pace of planet-hunting is accelerating. The number of known extrasolar worlds is almost certain to rise to thousands in the next few years. And if you think the planetary menagerie is crowded now, just wait until you hear about the oddballs that scientists are finding out in our galaxy's depths—including alien Plutos.

14

ALIEN PLUTOS

What if a world as small as Pluto were found in another planetary system? What if its orbit were as tangled up as those of the dwarf planets in our own solar system? Alan Boss, for one, would be absolutely thrilled. "We'd be happy to call it a planet candidate and just leave it at that," said the Carnegie Institution of Washington astronomer, who is a member of the International Astronomical Union committee focusing on alien planets and has served on the science team for NASA's Kepler planet-hunting mission.

The Kepler mission is just one of several scientific quests searching for alien planets, a quest that began in the 1990s. Most of the objects found so far are on the scale of Jupiter or Neptune. The Kepler space telescope and its European counterpart, known as COROT, are expected to identify smaller planets, down to the size of Earth itself. Next-generation telescopes would have to become still more powerful to spot alien Plutos—but some of the issues raised during the debate over dwarf planets are already starting to crop up in the wider search for worlds beyond our solar system.

It's easier for planet hunters to find larger objects because of the way the hunt is conducted. Until recently, the method most often used was to look for a pattern of ever-so-slight shifts in light caused by the motion of a star surrounded by planets. When a massive planet is orbiting a star, both bodies actually circle around a common center of gravity. For example, because of Jupiter's pull, our sun wobbles around a center of gravity at a speed of 12.5 meters per second, or about 28 miles per hour.[1] The wobble created by Earth is significantly less: 9 centimeters per second, or 0.2 miles per hour. That tiny shift makes it much harder to find Earth-sized planets—and well nigh impossible to find Pluto-sized planets.

Kepler and COROT (which is a crazy acronym standing for "COnvection, ROtation and planetary Transits") use a different method called transit photometry, which involves watching for the dip in starlight that takes place when a planet moves right across the star's disk. Between the two probes, hundreds of thousands of stars will be checked for evidence of planets, and the results are expected to lengthen the list of alien planets from mere hundreds to thousands.

The planet quest has shown that there's no shortage of diversity beyond our solar system. Astronomers have turned up plenty of worlds whirling around their parent stars in orbits much closer than Mercury's. Some of these blazing-hot planets are bigger than Jupiter; others are almost as small as Earth. Farther out from alien suns, other planets have been orbiting in a "Goldilocks zone" where the temperature is not too hot, not too cold, but just right for liquid water and life. Some stars are nestled in dusty cradles where planets are still in the process of forming. Other stars have blasted away at the planets orbiting them, leaving only burned-out cinders behind.

Although there's not yet an exact match for all Pluto's peculiarities, you can find an extrasolar system for virtually every characteristic that sets the dwarf planet apart from the solar system's mainstream.

For example, consider the first alien planetary system ever found, which was detected in 1991 around a rapidly spinning neutron star known as PSR B1257+12. Such neutron stars send out radio pulses on the scale of milliseconds, which makes it easier for scientists to calculate extremely slight perturbations due to their gravitational wobble. Astronomers Alexander Wolszczan and Dale Frail charted the orbits of two planets that were even less massive than Earth.

A third planet orbiting the pulsar is twice the mass of Earth's moon, and a fourth world appears to be smaller still, amounting to just one-fifth of Pluto's mass. Because of the uncertainties surrounding the fourth planet's detection, its existence still has to be fully confirmed—but so far, it sounds a lot like Pluto. "It is quite possible that the tiny fourth planet is the largest member of a cloud of interplanetary debris at the outer edge of the pulsar's planetary system," Wolszczan said.[2]

Other parallels to Pluto can be found within the disks of protoplanetary material swirling around alien stars. One study looked at the clumpiness of three such disks, around the stars AU Microscopii, Beta Pictoris, and Fomalhaut. The researchers behind the study suggested that Pluto-sized planetary embryos were forming inside all three of those disks.[3] Pluto-sized "baby planets" already may have formed in AU Microscopii's dusty womb, right about where Pluto would be in our own solar system.[4]

Fomalhaut, a bright star just twenty-five light-years distant from Earth, provides still more striking evidence of the planet-building process at work: Fomalhaut has an icy ring that is four times farther out than our own solar system's Kuiper Belt. The ring is off-kilter, suggesting to researchers that a Saturn-sized object was perturbing the material there.[5]

A couple of years after that initial finding, images from the Hubble Space Telescope revealed a planet-sized spot moving close to the edge of the ring—out much farther from Fomalhaut than Eris is from the sun today. The researchers behind the observations estimated the planet to be roughly Jupiter's mass.[6]

On the very same day, another research team reported spotting three planets around the star HR 8799, about 130 light-years from Earth. The closest-in planet of the three was just a few AU within where Neptune would be, and the others were more distant from their sun than Pluto is from ours.[7]

The two sets of findings marked the first time planets had been seen visually rather than detected indirectly. They also demonstrated that the outer parts of a star's neighborhood could be as ripe for planetary pickings as the core.

Would these planets qualify as "real" planets, or should they be counted as giant dwarf planets? The distinction didn't make much difference to Boss as he contemplated the groundbreaking Hubble image of Fomalhaut's planet. "That thing has not cleared its orbit yet, right?" he joked.

Dozens of extrasolar planets have been found in orbits that are far more eccentric than Pluto's. If a planet begins its life far from its parent star, it's less subject to the steady tug of that star's gravity, and more susceptible to the push and pull of its planetary neighbors. As a result, as you go farther out from the star, you're more likely to see orbits that are eccentric, inclined, or otherwise mixed-up.

Purists might have a hard time giving their planetary imprimatur to celestial objects that share an orbital zone with their neighbors, or worlds that weave in and out between other worlds. But these situations have occasionally cropped up in the extrasolar planet search—and we're likely to see more such curiosities as the search intensifies.

A classic example is the recently discovered planetary system around the star HD 45364, about 106 light-years from Earth. One planet is at least three times as massive as Neptune. Another is at least twice as massive as Saturn. The two planets trace eccentric orbits, veering inward and outward to such an extent that the "inner" planet can potentially stray outside the "outer" planet's path. And yet those orbits are projected to remain stable for five billion years. Why? Because they follow the same clockwork beat tapped out by Neptune and Pluto.

One planet completes three orbits in the same amount of time that the other planet takes to make two orbits. That two-to-three resonance means the larger planet can't possibly clear out the other, not quite so giant planet. "This is the first time that such an orbital resonant configuration has been observed for extrasolar planets, although an analogue does exist in our own solar system composed by Neptune and Pluto," the researchers reported.[8]

At least one other star—HD 82943, which is 89.5 light-years distant—has planets that go through an inner-outer switch similar to Neptune and Pluto.[9] Two other planets, circling a star called HD 128311, appear to trace even weirder contortions thanks to their two-to-one orbital clockwork.[10]

One of the closest stars known to have planets is Epsilon Eridani, a frequent setting for science-fiction tales. Although some *Star Trek* fans used to list it as the home star of Mr. Spock, one of television's most famous aliens, that honor has been shifted to a different star in the same constellation. Nevertheless, Spock would find Epsilon Eridani fascinating, and not just because it's on a list of stars most likely to have Earthlike planets.[11]

Astronomers say that the Epsilon Eridani planetary system appears to have an icy analog to our solar system's Kuiper Belt—in addition to not just one, but two asteroid belts.[12] And therein lies a puzzle: An earlier study had detected a planet with an orbit so eccentric that it would go right through the belt (or perhaps over and under it).[13] Because that runs counter to the orthodoxy about planetary formation, the astronomers behind the later study argued that the earlier measurements must have been wrong. It could take years to determine conclusively which study is correct—unless Spock himself decides to travel through a time-warping wormhole and fill us in.

These examples don't imply that there may be a planet orbiting a far-off star that's exactly like Pluto. On the contrary, the evidence hints that extrasolar planets are capable of behaving far more oddly than Pluto, Eris, Sedna, and all their eccentric cousins. The worlds on the edge of our solar system just might help us better understand the quirkier members of the universe's planetary menagerie.

The thousands of new planets scientists are likely to find will no doubt shed additional light on our own planetary family as well. In our solar system, there are wide gaps between the sizes of the giant planets (Jupiter and Saturn, Uranus and Neptune), the terrestrial planets (Earth, Venus, Mars, and Mercury), and the dwarf planets. Is there a universal reason behind that size breakdown, or is it merely a peculiarity of our own solar system? A broad census of the planetary spectrum may help answer that question.

The current methods for detecting planets aren't nearly sensitive enough to spot individual planetary bodies within the rocky or icy disks of distant stars, but studies of a nearby young star cluster have suggested that more than half of those star systems were in the process of building icy planets within huge rings of raw material. Ice dwarfs like Pluto may well serve as a typical template for planets in the making.[14]

When astronomers consider the diverse assortments of planetary systems that are likely to be found in the next few years, most of them would conclude that it's too early to get overly particular about what is and isn't formally considered a planet. The insights gained from Kepler, COROT, and other planet-hunting missions could well feed into a more sensible classification system for planets, modeled after the system that's used today to classify stars.[15]

About a century ago, astronomers Ejnar Hertzsprung and Henry Norris Russell drew up a diagram that charted the brightness of stars against their temperature and color. The result was a smooth "main sequence" of spectral classification, covering most types of stars as they evolved. The main-sequence categories range from O-type stars to M-type stars—a progression that astronomy students memorize with the phrase "Oh, Be A Fine Girl (or Guy), Kiss Me."[16] Other stars that don't fit the main sequence—such as giants and white dwarfs—clearly stick out on the chart as separate classes.

The Hertzsprung-Russell diagram and the main-sequence scale have served to organize the study of stars for decades. Would it be such a stretch to imagine a similar spectrum of main-sequence planets could be drawn up on a future chart—with ice dwarfs, rocky dwarfs, and other categories taking their rightful place as well?

Is it possible to come up with a unified planet theory, covering all the diversity we're finding beyond our own cosmic neighborhood? When the IAU began considering the questions surrounding planethood, some astronomers hoped that the resulting definition would cover planets beyond as well as within our solar system. That hope fell by the wayside during the Battle of Prague. After the battle's end, Geoff Marcy, a Berkeley astronomer and one of the world's foremost planet hunters, said trying to come up with a definition for planets beyond the solar system was a pointless exercise. "The taxonomy of asteroids, comets, moons, planets and brown dwarfs is far too limited to capture the diversity of their origins and internal constitutions," he said.[17]

Nevertheless, astronomers have applied a widely accepted rule of thumb to discriminate between planets and nonplanets, based on one physical characteristic: mass. The rule has nothing to do with where an object is, or how it is thought to have formed, or whether it's swept out its neighborhood. Here it is: Anything that is more than thirteen times as massive as Jupiter is judged not to be a planet, but rather a star or a brown dwarf—based on the calculation that the self-gravity of such objects is so powerfully crushing that it can light up a nuclear fusion reaction.

This rule of thumb defines the upper end of the planetary mass scale, but doesn't address the lower end. Astronomers have long suggested that mass could be used to define the lower end as well. However, they have disagreed over where to draw the line. Some wanted to draw it arbitrarily at Pluto's mass.[18] Others suggested drawing it at a tenth of Mercury's mass.[19] And in fact, the whole debate over physical roundness arose because such a standard would make more sense than an arbitrary cutoff.

"It is one constant thread that goes all the way through," Stern said. "It's the same physics that controls where an object becomes a planet and where it becomes a star."

Because of the way things turned out in Prague, extrasolar planet hunters are still left without a guideline for how small a planet could be. The IAU's planet definition explicitly addresses objects only in our own solar system, and nothing beyond it. That hasn't been a problem so far, but future discoveries could lead to some head-scratching.

"An Earth discovered around another star—say, by Kepler—would not necessarily be a planet if one extended the IAU definition to other solar systems," Mark Sykes explained. "The system might be too young, the distance of the Earth too far from the star, for instance. The IAU definition is meant to fit our solar system to get a specific result. It does not work well around other stars."

Is there a way to make things right? It took fifty years after Ceres's discovery to put the asteroid belt in some sort of perspective—and some astronomers still think Ceres hasn't gotten all the attention it deserves. More than seventy-five years after Pluto's discovery, astronomers are still debating the planet's proper place.

Getting just the right perspective on the thousands of planets yet to be discovered could take longer. Or maybe not all that long, if you're willing to adjust to a planetary paradigm shift.

15

THE CASE
FOR PLUTO

Never again can Pluto be the ninth planet. Or the littlest planet. Or the most distant planet. But does that make Pluto a nonplanet?

No way.

Even before Pluto was discovered, the solar system was divided into two classes of planets: the rocky worlds like Earth, and the gas giants beyond. Pluto has pointed the way to the solar system's third great class of planets, no less important than the other two.

Pluto isn't the ninth of nine; it's the first of many. Thanks to the discoveries of the past couple of decades, we've gained a whole new tribe of worlds to watch in the Kuiper Belt, and the vast, diffuse Oort Cloud represents an even farther frontier we haven't yet begun to explore.

These discoveries will shift our view of the universe the way Galileo and Copernicus shifted it four centuries ago. In the seventeenth century, the world came to understand that Earth was not the center of the universe. In the twenty-first century, we will come to understand that Earth provides just one template for the way the cosmos builds planets—and not even the most common template.

"The original view, until ten or fifteen years ago, was that we had four Earthlike terrestrial planets, four gas giants, and the misfit Pluto. But the new view is four terrestrial planets, four gas giants, and hundreds of Plutos," Alan Stern said. "It's jarring, because Pluto's no longer the misfit. It's the Earthlike planets that are the misfits."

Some people may find it difficult to handle a planetary paradigm shift, but shift happens, whether we like it or not.

For example, the conventional wisdom once held that very few planets existed in the universe. At the time Pluto was discovered, Sir Arthur Eddington estimated that only one star system out of 100 million had the right conditions to produce planets. Arthur Compton, one of the foremost American physicists of the day, declared that "a planet is a very rare occurrence."[1]

Eighty years later, the paradigm shift couldn't be more dramatic. The Geneva Observatory's Michel Mayor, a pioneer in the quest to find alien worlds, estimates that a third of all stars like the sun have planets ranging in size between Earth and Neptune.[2] What's more, planets are being found in formerly unthinkable places. One star harbors a super-Earth within a fraction of Mercury's distance from our sun. Another star has a Jupiter-scale planet that is three times farther away than Pluto is from the sun.[3]

The evidence emerging from the hunt for extrasolar planets would argue for going with a wide-ranging definition of planethood, based on physical properties. After all, that's how scientists define stars, ranging all the way from dwarfs to giants. Anything that's massive enough to fire up an internal fusion furnace is considered a star or a brown dwarf. Similarly, anything that's massive enough to build up layers of material into a gravitationally bound ball should be considered a planet.

A planet's ability to sweep out other objects in its vicinity is important, to be sure. When astronomers study how planetary systems are sculpted, they might find it useful to ignore the gravitational effects (or noneffects) of the individual smaller objects and think of them merely as parts of a larger population—say, a belt of asteroids, or a ring of comets. But that's no reason to draw an X through the legions of Planet Xs.

An overly narrow classification system is just asking to be rendered obsolete by future additions to the planetary list. Mike Brown, for instance, once suggested that Pluto and the dwarf planets he and his team discovered should not be considered true planets because they could be grouped with similar things. He said they should be seen merely as parts of a bigger population. "If you don't understand that, you get a misguided impression of the architecture of the solar system and how things formed and where things are," Brown said.[4]

But to paraphrase Shakespeare, there are more things in the heavens and on super-Earths than are dreamt of in that philosophy. The possibilities for planetary architecture are likely to be much more varied than we think. Even in our own solar system, we're discovering moons that lurk right in the middle of a planet's rings.[5] If that's so, then why can't planets lurk in the wide rings of material that surround our sun and other stars?

"What will Mike Brown say when we find a system with ten Saturns orbiting as a group?" Stern asked. "Or an Earth in the Oort Cloud? Or two Mercurys in close to one another? And what about when our own solar system was young, and Jupiter and Saturn crossed paths? Were they temporarily not planets during that era? Ridiculous, huh?"

To be sure, Stern isn't a dispassionate participant in this debate. He's a longtime Plutophile, as well as the principal investigator for New Horizons, NASA's mission to the dwarf planet and the Kuiper Belt. But a good number of planetary scientists who are watching from the sidelines agree with him that the International Astronomical Union's hastily written definition just won't cut it. And they'd like to see something done about that.

"If the IAU adopts a clearly flawed definition, nobody is under any obligation to accept it," said David Grinspoon, curator of astrobiology at the Denver Museum of Nature and Science. "But I'm getting sick of this. Do planetary scientists really want to be known as the community that can't stop fighting about what a planet is, during a decade when we are actually finding more planets every year than in all of human history, and launching spacecraft to solve mysteries of planetary climate, landscapes, and habitability?"

Grinspoon said it's time for a compromise that includes both a physical and a dynamical perspective. "*A planet is a round object orbiting a star*," he declared. "If we learn that it has not gravitationally dominated its surroundings, then it goes in a sub-class called dwarfs. Dwarf planets join terrestrial planets such as Earth and Jovian planets such as Jupiter or Gliese 581b as full-fledged citizens of their planetary systems, with all the rights and privileges thus implied."[6]

Strangely enough, it was just such a compromise that was voted down during the Battle of Prague.

It's worth asking whether all this hand-wringing over planethood really matters. Stern insists that agreeing on the meaning of the word is crucial to planetary scientists. "It's embarrassing to the field if we don't have a consensus," he said. "When a schoolkid or a schoolteacher or a person of the public who funds astronomy through their tax dollars says, 'Well, what's a planet,' we don't have a general consensus on what that means."

On the other side of the question, Neil deGrasse Tyson, who spent so much time deliberating over how planets should be presented at his museum, wonders whether the word has outlived its usefulness. "If I'm looking for a planet in another star system and I say, 'I just discovered a planet,' you then have to play twenty questions with me," he complained. "Is it big, is it small? Is it rocky, is it gaseous? Does it have rings, does it have moons? Is it close, is it far? Is it in a habitable zone? Might it have water? And so once I told you I discovered a planet, the word has no utility anymore."[7]

Tyson's rant may be rhetorical, but it also contains a grain of truth: One word is no longer enough to provide the full picture of planethood. Today, the term covers a wide spectrum of worlds, and that spectrum is certain to get wider as more discoveries are made. That doesn't mean the word itself is spoiled. Scientists in other disciplines face similar issues with other words, such as "organism" and "particle." It's only natural that a concept so fundamental to an entire field of science should be so broad.

In such a situation, adjectives and qualifiers can come in handy: dwarf versus dominant, scattered versus classical, Earthlike versus giant versus icy. Such are the classifications that will emerge as the study of planets comes into its own.

"It is the promise of comparative planetology that drives most investigations today," Mark Sykes said. "How do processes work and manifest themselves on bodies of different composition, masses, and stellar distance? This gives us fundamental new insights into how those processes work on our own planet."

Admitting the dwarfs as members of the planetary club will pay dividends during the next generation of exploration, according to Gibor Basri, an astronomer at the University of California at Berkeley. "I think it's actually exciting that we can find more planets in our own solar system," Basri said. "I think it gets kids more interested ... maybe they could grow up and find another planet in our solar system."

And as far as Basri is concerned, it's still perfectly fine to make a distinction between the solar system's eight biggest planets and the rest. Making distinctions is the whole point of planetology. "It's clear that our solar system has eight major planets and then a number of dwarf planets," he said. "That's okay with me, as long as it doesn't kick them out of the planet category."[8]

By now, it should be obvious that the case for Pluto isn't just about one picked-upon planet. On one level, this case is a class-action suit—affecting the status of other worlds in our solar system, and potentially thousands of worlds beyond. On a deeper level, it's a case study that shows how politics and personalities can affect the scientific process, and how the scientific process can in turn affect popular culture.

The IAU issued its ruling in the case back in 2006, but since then it has been on appeal. So who's the ultimate judge?

Thankfully, scientific matters aren't decided by a single vote. "We figured out that water was once prevalent on Mars, not by getting together and arbitrarily calling committees to vote on it, but because over time the body of evidence became overwhelming," Stern pointed out.

Sorting through a body of scientific evidence can take decades. That was certainly the case for the theory of continental drift. It took centuries to develop a hypothesis about the movements of Earth's continental plates, and four decades to work out the mechanism behind those movements. Along the way, geologists argued over the evidence and cast doubts on the theories—and sometimes the debate got downright personal.

"Science is a lot about disagreements and mud fights," Sykes said.

Ultimately, it's up to the scientific community and the general public to decide how planets will be classified. Sometimes those two constituencies will go in different directions. For example, ask a botanist whether tomatoes, corn, and green beans are vegetables. Then go ask a cook.

No matter which label scientists try to attach—dwarf planets or minor planets, Kuiper Belt objects or iceballs—Pluto and its far-flung cousins are well worth our wonder. And the best is yet to come.

A dwarf-planet extravaganza is due to begin in 2011 when NASA's Dawn spacecraft flies around the asteroid Vesta, the brightest object in the asteroid belt. Some astronomers think Vesta might qualify as a dwarf planet, even though a huge chunk was blasted off the space rock millions of years ago in a cosmic collision. About 6 percent of all the meteorites that fall to Earth are thought to be bits of debris from that blast. Whether Vesta ends up being called a planet or not, Dawn will revolutionize our view of that broken world.

After the rendezvous with Vesta, Dawn zooms on to Ceres. When the spacecraft goes into orbit around the dwarf planet in 2015, mission scientists hope to watch clouds sailing through Ceres's thin atmosphere, study its clay-rich surface, and look for evidence of water lying beneath the surface.

That same year, NASA's New Horizons spacecraft will finally reach Pluto and Charon after a nine-year cruise. Thousands of pictures of the ice dwarfs will stream back to Earth, perhaps revealing ice volcanoes, wide stretches of methane frost, and hydrocarbon mud. After the Pluto-Charon flyby, the mission team will try to aim New Horizons toward one or two other icy denizens of the Kuiper Belt.

Stern said the space mission could provide "a window four and a half billion years back in time" to learn how the solar system's larger planets were formed.

Vanderbilt University's David Weintraub said the Dawn and New Horizons missions could well give Ceres and Pluto a fresh boost of positive press. "They certainly will take on a new life when we see them as real objects," he said. "I think that will continue the love affair with Pluto—and I think it will enhance the status of Ceres, too."

So don't count Pluto out yet. This case is far from closed.

APPENDIX A

What to Tell Your Kids about Planets

Kids and parents alike are often confused by the controversies over Pluto and Eris, planets and dwarf planets, clearing out neighborhoods and achieving hydrostatic equilibrium. Here are some straightforward answers to eleven commonly asked questions:

What's a planet?

All planets are huge balls in outer space that can contain rock, gas, or ice. Our Earth is a great example of a planet. Some planets, like Jupiter, are much bigger. Other planets, like Pluto, are much smaller. Scientists are even finding planets orbiting other stars, hundreds of trillions of miles away.

How many planets go around the sun?

Right now there are four, plus four, plus more.

Four planets—Mercury, Venus, Earth, and Mars—are made almost entirely of rock. We call these terrestrial planets because they're like Earth, which was known as *Terra* to the Romans.

Another four planets—Jupiter, Saturn, Uranus, and Neptune—are giants that are topped with thick atmospheres. Unlike terrestrial planets, these giants don't have a solid surface you could land a spaceship on. Jupiter and Saturn are mostly made of hydrogen and helium, which are the elements found in the sun. Uranus and Neptune are mostly made of different types of ice, including water and methane ice.

There are more planets that are smaller than the giants and the terrestrial planets. Some people call these dwarf planets. One of them, Ceres, is part of the asteroid belt between Mars and Jupiter. Pluto was the first planet found beyond Neptune and is also a dwarf. Telescopes have gotten a lot more powerful since Pluto was discovered. Scientists are using those telescopes to find more dwarf planets that are even farther away than Pluto.

There could be larger planets waiting to be discovered, way out in the solar system. Maybe you can be the first person to find some of them when you grow up!

Isn't Pluto too small to be a planet?

No. The bigger an object is, the more gravity it has, and things on its surface get heavier. If the object is big enough, gravity crushes its rock and ice into a huge ball—and that's what makes it a planet. Scientists have confirmed that Pluto is a big ball, based on years of telescope observations.

Pluto has an atmosphere and three moons, as well as seasons and geology—and these are all things that scientists think about when they think about planets. Pluto is smaller than Earth, but it's much bigger than the smallest planet we know of, which is Ceres.

Do all scientists think the same way about planets?

No. Some scientists think a planet has to be all by itself and push almost everything else out of its way. Because Pluto shares a wide area of outer space with lots of other, smaller objects, they don't think it should be called a planet. They also don't count Ceres or the other dwarfs as "real" planets.

Other scientists, however, think it's okay for a planet to share its space. Scientists are continuing to debate the issue, and as they learn more about planets, some of them change their minds. It's not a bad thing that scientists sometimes disagree. That's how science works: New discoveries lead to different ways of looking at the universe, and it can take years to decide which way is best.

What happens if an object in space isn't big enough to have a round shape?

If it's mostly icy, it could be a comet—particularly when it comes close to the sun and gives off a bright tail of streaming gas and dust. If it's mostly rocky, it could be an asteroid. Asteroids can't have a lot of the things that planets can have, like an atmosphere, but they can have moons going around them. Asteroids and comets can look like potatoes, or dog bones, or gigantic pebbles. They're just not big enough for gravity to crush them into a nice round shape.

Can a moon be a planet?

Usually we think of a moon as anything that goes around a planet, and we wouldn't consider that to be a planet itself. But moons can be big and round, just like planets. In fact, some moons are bigger than some planets. A moon that goes around Saturn, called Titan, is an excellent example. It has rivers that are made of cold liquid methane and a thick, hazy atmosphere. A moon that goes around Neptune, called Triton, might have been a planet billions of years ago but was captured in orbit by the bigger planet—and that's why we call Triton a moon rather than a planet today. Other moons, like the two moons that go around Mars, are too small to be planets. They may have been asteroids that were captured in Martian orbit.

Some scientists think that if a moon is big enough to be round, it should also be considered a planet, or at least a planetary object—because if you were to fly a spaceship there and look out the window, it would look like a planet that orbits the sun.

How can I remember the names of the planets in our solar system?

You don't need to memorize the names of all the planets, just like you don't need to memorize the names of all the dinosaurs, or all the world's rivers. Your teacher might want you to remember the biggest planets, but never forget that there's more to the solar system than just the four, eight, or thirteen biggest things. Asteroids, comets, and moons can be just as interesting as planets.

If you want to remember the "four plus four" planets, here's a sentence that helps you do it: "My Very Excellent Mother Just Served Us Nachos." The first letters of the first four words stand for the four terrestrial planets, in order from the sun: Mercury, Venus, Earth, and Mars. The second four initials stand for the solar system's giants: Jupiter, Saturn, Uranus, and Neptune.

As you get older, you expand your diet to include more sophisticated foods, and not just nachos. The same goes for the planetary menu. One way to add to your "diet" is to include Ceres, the first rocky dwarf planet found in the asteroid belt; and Pluto, the first icy dwarf planet found in the Kuiper Belt. With those additions in mind, here's a more sophisticated memory aid: "My Very Excellent Master Chef Just Served Us New Potatoes."

Another fun thing to do is to memorize the first five recognized dwarf planets, using this comment about your very excellent mother: "Extra Planets Make Her Crazy." The initials stand for Eris, Pluto, Makemake, Haumea, and Ceres.

Can something be too big to be a planet?

Yes. Let's take Jupiter as an example. That giant planet is made mostly of hydrogen, just like the sun. If Jupiter had been able to get a lot bigger, it could have become another star like the sun. Stars are bright because they are burning their hydrogen in a process called nuclear fusion. Scientists have figured out that fusion begins to happen inside an object when it is thirteen times as massive as Jupiter.

When an object gets that big, it's called a brown dwarf or a star, depending on its size.

Does a planet have to go around the sun or another star?
That's another point that scientists are debating. Some of them have reported detecting "free-floaters," or rogue planets, out where there are no stars. Other scientists think these rogue planets are actually more like brown dwarfs. It's a mystery you might be able to help solve when you grow up.

How many planets are there beyond our solar system?
There are certainly thousands if not millions or billions of planets that orbit other stars in our galaxy. So far, scientists have found hundreds of such planets, and some of them are unlike anything we've seen in own solar system. There are planets that whirl around their suns in incredibly close orbits. Some of these "hot planets" are bigger than Jupiter, while others are nearly as small as Earth. Other planets follow orbits more eccentric than any of the planets in our solar system. Temperatures on those planets can swing between boiling hot and freezing cold. Still other planets are in orbits that could make them just right to live on—not too hot, and not too cold.

Today, our telescopes aren't powerful enough to spot planets that are Earth's size or smaller, but they should be able to detect alien Earths in the next few years.

Are there things living on other planets?
No one has yet found clear evidence of life beyond Earth, but it's possible. Some scientists think that simple forms of

life—perhaps like germs—could live beneath the surface of planets or moons in our own solar system. Looking beyond our solar system, scientists are trying to use telescope readings to find out whether distant planets could support life.

Even if we detect life on a planet orbiting another star, the distance between stars is so vast that we know of no practical way to travel back and forth. We couldn't visit them, and they couldn't visit us. But no matter how far away it is, or how small it is, any planet that has living things would be a Very Important Planet.

APPENDIX B

The International Astronomical Union's Resolutions and Revisions

Here are the texts for the original draft resolution on planet definition, proposed to the International Astronomical Union on August 16, 2006; and the revised resolutions as voted upon on August 24, 2006.

Draft Resolution 5: Definition of a Planet (August 16, 2006)

Contemporary observations are changing our understanding of the Solar System, and it is important that our nomenclature for objects reflect our current understanding. This applies, in particular, to the designation "planets." The word "planet" originally described "wanderers" that were known only as moving lights in the sky. Recent discoveries force us to create a new definition, which we can make using currently available scientific information. (Here we are not concerned with the upper boundary between "planet" and "star.")

The IAU therefore resolves that planets and other Solar System bodies be defined in the following way:

1. A planet is a celestial body that (a) has sufficient mass for its self-gravity to overcome rigid body forces so that it assumes a hydrostatic equilibrium (nearly round) shape[1], and (b) is in orbit around a star, and is neither a star nor a satellite of a planet.[2]

2. We distinguish between the eight classical planets discovered before 1900, which move in nearly circular orbits close to the ecliptic plane, and other planetary objects in orbit around the Sun. All of these objects are smaller than Mercury. We recognize that Ceres is a planet by the above scientific definition. For historical reasons, one may choose to distinguish Ceres from the classical planets by referring to it as a "dwarf planet."[3]

3. We recognize Pluto to be a planet by the above scientific definition, as are one or more recently discovered large Trans-Neptunian Objects. In contrast to the classical planets, these objects typically have highly inclined orbits with large eccentricities and orbital periods in excess of 200 years. We designate this category of planetary objects, of which Pluto is the prototype, as a new class that we call "plutons."

4. All non-planet objects orbiting the Sun shall be referred to collectively as "Small Solar System Bodies."[4]

Revised Resolutions (August 24, 2006)

Resolution 5A is the principal definition for the IAU usage of "planet" and related terms. Resolution 5B adds the word "classical" to the collective name of the eight planets Mercury through Neptune.

Resolution 6A creates for IAU usage a new class of objects, for which Pluto is the prototype. Resolution 6B introduces the name "plutonian objects" for this class. The Merriam-Webster dictionary defines "plutonian" as:

> Main Entry: plu·to·ni·an
> Pronunciation: plü-'tō-nē- ən
> Function: *adjective*
> Usage: *often capitalized*
> of, relating to, or characteristic of Pluto or the lower world

After having received inputs from many sides—especially the geological community—the term "Pluton" is no longer being considered.

IAU Resolution: Definition of a "Planet" in the Solar System

Contemporary observations are changing our understanding of planetary systems, and it is important that our nomenclature for objects reflect our current understanding. This applies, in particular, to the designation "planets." The word "planet" originally described "wanderers" that were known only as moving lights in the sky. Recent discoveries lead us to create a new definition, which we can make using currently available scientific information.

Resolution 5A (Approved: without recorded vote)

The IAU therefore resolves that planets and other bodies in our Solar System, except satellites, be defined into three distinct categories in the following way:

1. A "planet"[1] is a celestial body that (a) is in orbit around the Sun, (b) has sufficient mass for its self-gravity to overcome rigid body forces so that it assumes a hydrostatic equilibrium (nearly round) shape, and (c) has cleared the neighborhood around its orbit.

2. A "dwarf planet" is a celestial body that (a) is in orbit around the Sun, (b) has sufficient mass for its self-gravity to overcome rigid body forces so that it assumes a hydrostatic equilibrium (nearly round) shape,[2] (c) has not cleared the neighbourhood around its orbit, and (d) is not a satellite.

3. All other objects,[3] except satellites, orbiting the Sun shall be referred to collectively as "Small Solar System Bodies."

Resolution 5B (Rejected: 91 votes in favor; more were against, no count taken)

Insert the word "classical" before the word "planet" in Resolution 5A, Section (1), and footnote 1. Thus reading:

(1) A classical "planet"[4] is a celestial body . . .

IAU Resolution: Pluto

Resolution 6A (Approved: 237 in favor, 157 against, 17 abstaining)

The IAU further resolves:

Pluto is a "dwarf planet" by the above definition and is recognized as the prototype of a new category of trans-Neptunian objects.

Resolution 6B (Rejected: 183 in favor, 186 against)

The following sentence is added to Resolution 6A:

This category is to be called "plutonian objects."

Footnotes for the August 16 Resolution

1 This generally applies to objects with mass above 5×10^{20} kg and diameter greater than 800 km. An IAU process will be established to evaluate planet candidates near this boundary.

2 For two or more objects comprising a multiple object system, the primary object is designated a planet if it independently satisfies the conditions above. A secondary object satisfying these conditions is also designated a planet if the system barycenter resides outside the primary. Secondary objects not satisfying these criteria are "satellites." Under this definition, Pluto's companion Charon is a planet, making Pluto-Charon a double planet.

3 If Pallas, Vesta, and/or Hygeia are found to be in hydrostatic equilibrium, they are also planets, and may be referred to as "dwarf planets."

4 This class currently includes most of the Solar System asteroids, near-Earth objects (NEOs), Mars-, Jupiter-, and Neptune-Trojan asteroids, most Centaurs, most Trans-Neptunian Objects (TNOs), and comets. In the new nomenclature the concept "minor planet" is not used.

Footnotes for the August 24 Resolution

1 The eight planets are: Mercury, Venus, Earth, Mars, Jupiter, Saturn, Uranus, and Neptune.

2 An IAU process will be established to assign borderline objects into either "dwarf planet" and other categories.

3 These currently include most of the Solar System asteroids, most Trans-Neptunian Objects (TNOs), comets, and other small bodies.

4 The eight classical planets are: Mercury, Venus, Earth, Mars, Jupiter, Saturn, Uranus, and Neptune.

APPENDIX C

Planetary Vital Statistics

Here's a quick rundown on the terrestrial and the giant planets of our solar system, plus the first five objects to be recognized as dwarf planets by the International Astronomical Union. AU stands for astronomical units. One AU is the distance from Earth to the sun (93 million miles, or 149.6 million kilometers):

Terrestrial planets

Mercury
   Mass (Earth = 1): 0.055
   Equatorial diameter: 3,031 miles (4,878 kilometers)
   Mean density (water = 1): 5.43
   Orbital distance from sun: 0.31 to 0.46 AU
   Mercury year = 88 Earth days
   Mercury solar day = 176 Earth days
   Moons: None
Venus
   Mass (Earth = 1): 0.81

Equatorial diameter: 7,521 miles (12,104 kilometers)
Mean density (water = 1): 5.24
Orbital distance from sun: 0.72 to 0.73 AU
Venus year = 224.7 Earth days
Venus solar day = 117 Earth days
Moons: None

Earth

Mass (Earth = 1): 1
Equatorial diameter: 7,926 miles (12,756 kilometers)
Mean density (water = 1): 5.52
Orbital distance from sun: 1 AU
Earth year = 365.25 Earth days
Earth day = 24 hours
Moons: *Luna* ("the moon")

Mars

Mass (Earth = 1): 0.11
Equatorial diameter: 4,217 miles (6,787 kilometers)
Mean density (water = 1): 3.94
Orbital distance from sun: 1.4 to 1.7 AU
Mars year = 687 Earth days
Mars day = 24 hours, 37 minutes
Moons: Phobos and Deimos

Giant planets

Jupiter

Mass (Earth = 1): 317.94
Equatorial diameter: 88,700 miles (142,800
kilometers)

Mean density (water = 1): 1.33
Orbital distance from sun: 5.0 to 5.5 AU
Jupiter year = 11.857 Earth years
Jupiter day = 9 hours, 56 minutes
Moons: At least 63

Saturn

Mass (Earth = 1): 95.159
Equatorial diameter: 74,600 miles (120,000 kilometers)
Mean density (water = 1): 0.7
Orbital distance from sun: 9.0 to 10.1 AU
Saturn year = 29.5 Earth years
Saturn day = 10 hours, 39 minutes
Moons: At least 61

Uranus

Mass (Earth = 1): 14.5
Equatorial diameter: 31,800 miles (51,200 kilometers)
Mean density (water = 1): 1.3
Orbital distance from sun: 18.4 to 20.1 AU
Uranus year = 83.75 Earth years
Uranus day = 0.72 Earth days
Moons: 27

Neptune

Mass (Earth = 1): 17.131
Equatorial diameter: 30,770 miles (49,520 kilometers)
Mean density (water = 1): 1.64
Orbital distance from sun: 29.8 to 30.4 AU
Neptune year = 163.73 Earth years

Neptune day = 0.671 Earth days
Moons: 13

Dwarf planets

Ceres

Mass (Earth = 1): 0.00016

Equatorial diameter: 606 miles (974.6 kilometers)

Mean density (water = 1): 2.1

Orbital distance from sun: 2.55 to 2.99 AU

Ceres year = 4.6 Earth years

Ceres day = 9 hours

Moons: None

Pluto

Mass (Earth = 1): 0.002

Equatorial diameter: 1,430 miles (2,302 kilometers)

Mean density (water = 1): 2.0

Orbital distance from sun: 29.7 to 49.3 AU

Pluto year = 248 Earth years

Pluto day = 6.4 Earth days

Moons: Charon, Nix, and Hydra

Makemake

Mass (Pluto = 1): 0.3

Diameter: 800–1,200 miles (1,300–1,900 kilometers)

Mean density (water = 1): 2.0?

Orbital distance from sun: 38.5 to 53.1 AU

Makemake year = 309.9 Earth years

Makemake day = not yet known

Moons: None

Haumea

    Mass (Pluto = 1): 0.32

    Diameter: 1,218 by 943 by 619 miles (1,960 by 1,518 by 996 kilometers)

    Mean density (water = 1): 2.6 to 3.3

    Orbital distance from sun: 34.7 to 51.5 AU

    Haumea year = 285 Earth years

    Haumea day = 3 hours, 55 minutes

    Moons: Namaka and Hi'iaka

Eris

    Mass (Pluto = 1): 1.27

    Diameter: 1,500 miles (2,400 kilometers)

    Mean density (water = 1): 1.18 to 2.31

    Orbital distance from sun: 37.8 to 97.6 AU

    Eris year = 557 Earth years

    Eris day = 1.08 Earth days

    Moons: Dysnomia

And more planets are sure to be added to the list. (Sources include NASA, SolarViews.com, the Planetary Society, *Windows to the Universe* [http://www.windows.ucar.edu/], Wikipedia, *Icarus* and *Planets Beyond: Discovering the Outer Solar System*, by Mark Littmann.)

NOTES

1. The Planet in the Cornfield

1 Alan Stern quoted in Paul Rincon, "Pluto Vote 'Hijacked' in Revolt," BBC News, August 25, 2006, http://news.bbc.co.uk/2/hi/science/nature/5283956.stm, accessed May 27, 2009.

2 Michael Brown, "The Great Planet Debate Wasn't," *Mike Brown's Planets,* August 17, 2008, http://www.mikebrownsplanets.com/2008/08/great-planet-debate-wasnt.html, accessed May 27, 2009.

3 Comments posted to Alan Boyle's blog, *Cosmic Log,* "The Lighter Side of Pluto," August 24, 2006, http://cosmiclog.msnbc.msn.com/archive/2006/08/24/2410.aspx, accessed May 27, 2009.

4 Quoted in Robert Roy Britt, "Pluto Demoted: No Longer a Planet in Highly Controversial Definition," Space.com, August 24, 2006, http://www.space.com/scienceastronomy/060824_planet_definition.html, accessed May 27, 2009.

5 HL Tau b, announced at the Royal Astronomical Society's National Astronomy Meeting, April 1, 2008.

6 The Epsilon Eridani system, accepted for publication in the *Astronomical Journal,* January 10, 2009.

7 Quoted in Drew Sefton, "Kansas Lad's 'Planet' Soon May Be No More," *Kansas City Star,* January 22, 1999, F1.

8 Frank Deford, "Pluto's Lessons for the World of American Sports," *Morning Edition,* National Public Radio, August 30, 2006, http://www.npr.org/templates/story/story.php?storyId=5736035, accessed May 27, 2009.

9 Andy Borowitz, "Scientists Demote Bush Presidency to Dwarf Status," *The Borowitz Report,* November 15, 2006, http://www.borowitzreport .com/article.aspx?ID=6632, accessed May 27, 2009.

2. Fellow Wanderers

1 Fred William Price, *The Planet Observer's Handbook* (Cambridge, UK: Cambridge University Press, 2000), 27.

2 Galileo Galilei, *Sidereus Nuncius* (*The Sidereal Messenger*), published in Italian in 1610 and translated into English by Albert Helden (Chicago: University of Chicago Press, 1989).

3 Not just one but two popes tried to make amends centuries after the fact. In 1991, John Paul II said Galileo's punishment was a mistake resulting from a "tragic, mutual miscomprehension." And in 2008, Benedict XVI paid tribute to Galileo as a scientist who helped the faithful "contemplate with gratitude the Lord's works." Quoted in Nicole Winfield, "Good Heavens: Vatican Rehabilitates Galileo," Associated Press dispatch, December 23, 2008, http://www.msnbc.msn.com/id/28371902/, accessed May 27, 2009.

4 David A. Weintraub, *Is Pluto a Planet? A Historical Journey Through the Solar System* (Princeton, NJ: Princeton University Press, 2007), 65.

5 Michael Lemonick, *The Georgian Star: How William and Caroline Herschel Revolutionized Our Understanding of the Cosmos* (New York: W.W. Norton & Co., 2008), 77.

6 Tom Standage, *The Neptune File: A Story of Astronomical Rivalry and the Pioneers of Planet Hunting* (New York: Walker & Co, 2000), 17.

7 Quoted in Lemonick, *The Georgian Star*, 141.

8 Lemonick, *The Georgian Star*, 144–145.

9 Standage, *The Neptune File*, 39.

10 Jeff Schnaufer, "Scientists Squabble over Rock Named Spock," *Los Angeles Times*, July 23, 1994, B1.

11 Alan Boyle, "Asteroid Named After 'Hitchhiker' Humorist," MSNBC. com, January 25, 2005, http://www.msnbc.msn.com/id/6867061, accessed May 27, 2009.

12 Gerhard Wurm and Jurgen Blum, "Experiments on Planetesimal Formation," in *Planet Formation: Theory, Observation and Experiments,*

eds. Herbert Klahr and Wolfgang Brandner (Cambridge, UK: Cambridge University Press, 2006), 90–111.

13 P. C. Thomas et al., "Differentiation of the Asteroid Ceres as Revealed by Its Shape," *Nature* 437 (September 8, 2005): 224–226.

14 James L. Hilton, "When Did the Asteroids Become Minor Planets?" Astronomical Applications Department, U.S. Naval Observatory, 2006, http://aa.usno.navy.mil/faq/docs/minorplanets.php, accessed May 27, 2009.

15 Mark Littmann, *Planets Beyond: Discovering the Outer Solar System* (New York: John Wiley & Sons, 1990), 28.

16 Standage, *The Neptune File*, 114–122. Emphasis in the original.

17 Quoted in Denison Olmsted, "Thoughts on the Discovery of Le Verrier's Planet," *New York Municipal Gazette* 1, no. 46 (February 1, 1847): 670.

18 David W. Hughes, "Galileo Saw Neptune in 1612," *Nature* 287 (September 25, 1980): 277–278.

3. The Search for Planet X

1 Other such millionaires included Charles Yerkes (Yerkes Observatory), James Lick (Lick Observatory), and Andrew Carnegie (Mount Wilson and Palomar observatories). Another grand generation of millionaires and billionaires is at work today, including Paul Allen and Nathan Myhrvold (Allen Telescope Array), as well as Charles Simonyi and Bill Gates (Large Synoptic Survey Telescope).

2 David Strauss, *Percival Lowell: The Culture and Science of a Boston Brahmin* (Cambridge, MA: Harvard University Press, 2001), 55.

3 Quoted in David H. Levy, *Clyde Tombaugh: Discoverer of Planet Pluto* (Tucson: University of Arizona Press, 1991), 101.

4 David A. Weintraub, *Is Pluto a Planet? A Historical Journey Through the Solar System* (Princeton, NJ: Princeton University Press, 2007), 125.

5 Walter Isaacson, *Einstein: His Life and Universe* (New York: Simon & Schuster, 2007), 199–223.

6 Strauss, *Percival Lowell*, 253.

7 E. M. Standish, "Planet X—No Dynamical Evidence in the Optical Observations," *Astronomical Journal* 105, no. 5 (May 1993): 2000–2006.

8 Nancy Zaroulis, "The Man Who Invented Mars," *Boston Globe Sunday Magazine*, April 27, 2008, 28.

9 William Lowell Putnam, *Percival Lowell's Big Red Car: The Tale of an Astronomer and a 1911 Stevens-Duryea* (Jefferson, NC: McFarland & Co., 2002), 150.

10 Clyde W. Tombaugh and Patrick Moore, *Out of the Darkness: The Planet Pluto* (Harrisburg, PA: Stackpole Books, 1980), 117.

11 Levy, *Clyde Tombaugh*, 55.

12 Ibid., 1–7.

13 Information Please ranks the Tigris at the bottom of its list of the world's fifty-four principal rivers, based on length (http://www.infoplease .com/ipa/A0001779.html, accessed May 28, 2009). And some scholars say the mountain known today as Mount Sinai is so unimpressive that they suspect the Bible was actually referring to a different peak: Mount Catherine, Egypt's highest mountain.

4. Pluto and Its Little Pals

1 Clyde W. Tombaugh, *The Search for the Ninth Planet, Pluto*, Astronomical Society of the Pacific Leaflets, Leaflet No. 209 (July 1946), 74.

2 David Levy, *Clyde Tombaugh: Discoverer of Planet Pluto* (Tucson: University of Arizona Press, 1991), 65.

3 Ibid., 69–70.

4 Associated Press, "Ninth Planet Discovered on Edge of Solar System; First Found in 84 Years," in *New York Times*, March 14, 1930, 1.

5 Mark Littmann, *Planets Beyond: Discovering the Outer Solar System* (New York: John Wiley & Sons, 1990), 84.

6 Levy, *Clyde Tombaugh*, 78. Pluto Water's slogan can be found in ads available online via the *Ad Art Gallery*, http://www.gono.com/adart/ adartgallery.php, accessed May 28, 2009.

7 Pluto had been suggested as a possible name for Planet X as far back as 1919 (Alan Stern and Jacqueline Mitton, *Pluto and Charon: Ice Worlds on the Ragged Edge of the Solar System*, 2nd ed. [Weinheim, Germany: Wiley-VCH, 2005], 235), and Tombaugh said he thought the name was on the Lowell Observatory's internal list a couple of weeks before he heard that Burney had suggested it. (Levy, *Clyde Tombaugh*, 78).

8 Edward Goldstein, "The Girl Who Named a Planet: Interview with Venetia Burney Phair," NASA, January 17, 2006, http://www.nasa .gov/multimedia/podcasting/transcript_pluto_naming_podcast.html, accessed May 28, 2009.

9 Govert Schilling, *The Hunt for Planet X: New Worlds and the Fate of Pluto* (New York: Copernicus Books, 2009), 43.

10 Paul Rincon, "The Girl Who Named a Planet," BBC News, January 13, 2006, http://news.bbc.co.uk/1/hi/sci/tech/4596246.stm, accessed May 28, 2009.

11 William Graves Hoyt, "W. H. Pickering's Planetary Predictions and the Discovery of Pluto," *Isis* 67, no. 4 (December 1976): 563–564.

12 Littmann, *Planets Beyond*, p. 85.

13 Quoted in "Here's Pluto!" in *Tammy's Little Place of Disney*, http://www.geocities.com/misspcgenius/library/pluto.html, accessed May 28, 2009.

14 Interview with Dave Smith, founder and chief archivist of the Walt Disney Archives, February 11, 2009. By the way, Pluto also inspired racier tales as well. One of the first sci-fi stories to feature Pluto was titled "Into Plutonian Depths," published in *Wonder Stories Quarterly* in 1931. In this sexual satire, there are three sexes on Pluto, and according to a description of the story, the two visiting Earthmen "become very friendly with Zandaye, an aberrant Plutonian woman who is considered a monster of fatness and odd coloration by the Plutonians, but approaches a desirable human type." Of course, the hero falls in love with Zandaye, and the rest is pulp fiction. (Everett Franklin Bleiler and Richard Bleiler, *Science-Fiction: The Gernsbach Years: A Complete Coverage of the Genre Magazines from 1926 through 1936*, [Kent, OH: Kent State University Press, 1998], 71)

15 For the sake of convenience, future references to orbital distances will be expressed in terms of AU.

16 The 24-inch telescope used to record that spectrum was donated years later to the Eastern Iowa Observatory and Learning Center, a facility that my friend Chief and other astronomy club members built just outside Mount Vernon, Iowa.

5. The Meaning of a Moon

1 James W. Christy, "A Moment of Perception," in Mark Littmann, *Planets Beyond: Discovering the Outer Solar System* (New York: John Wiley & Sons, 1990), 175.

2 Alan Stern and Jacqueline Mitton, *Pluto and Charon: Ice Worlds on the Ragged Edge of the Solar System*, 2nd ed. (Weinheim, Germany: Wiley-VCH, 2005), 49–55.

3 Christy, "A Moment of Perception," 176.

4 Pluto's seasonal temperatures are thought to vary from –360 degrees Fahrenheit (–218 degrees Celsius or 55 Kelvin) to –400 degrees Fahrenheit (–240 degrees Celsius or 33 Kelvin).

6. There Goes the Neighborhood

1 In 1977, astronomer Carl Sagan observed that Chiron was part of a "horde of little planets" orbiting between Jupiter and Neptune. Such objects are today known as centaur asteroids. Unlike Pluto, centaurs follow unstable orbits and likely last only a few million years before running into something else.

2 B. G. Marsden, "Planets and Satellites Galore," *Icarus* 44 (1980): 32.

3 The name and number for (330) Adalberta were given to a different asteroid in 1982, but Marsden's jest was still taken seriously—so seriously that it turned up in the *New York Times* in 1987 (Irving Molotsky, "Pluto Facing Indentity [*sic*] Crisis: Is It a Planet or an Asteroid?" March 30, 1987, B6). Marsden had to write a letter to the editor insisting that his suggestion for renaming Pluto was a joke (Brian G. Marsden, "Let's Not Reclassify Pluto Until We Know Better What It Is," Letter to the Editor, *New York Times*, May 6, 1987, A34).

4 Pluto's orbit is more inclined and eccentric than those of the solar system's largest planets—but not too far from the inclination and eccentricity of Mercury's orbit (which has a seven-degree inclination and 0.21 eccentricity, compared with Pluto's 17 degrees and 0.25 eccentricity).

5 Lucy Ann Adams McFadden, Paul Robert Weissman, and Torrence V. Johnson, eds., *Encyclopedia of the Solar System*, 2nd ed. (San Diego: Academic Press, 2007), 543. Even the orbital paths of Pluto and Neptune come no closer than 2.4 AU. In comparison, the minimum Earth-Mars separation is roughly 0.37 AU, and Mars's mean distance from the sun is 1.52 AU.

6 John Davies, *Beyond Pluto: Exploring the Outer Limits of the Solar System* (Cambridge, UK: Cambridge University Press, 2001), 94–95.

7 Renu Malhotra, "Migrating Planets," *Scientific American* 281, no. 3 (September 1999): 56–63. Computer simulations conducted in the 1980s indicated that Pluto's motion is chaotic over a time scale of twenty million years (Gerald Jay Sussman and Jack Wisdom, "Numerical Evidence That the Motion of Pluto Is Chaotic," *Science*, July 22, 1988, 433), but then again, the inner solar system's motions are chaotic over a scale of

four to five million years (Wayne B. Hayes, "Is the Outer Solar System Chaotic?" *Nature Physics* 3, 689, published online in advance of print September 23, 2007).

8 History has been kind to the Dutch when it comes to ascribing credit for these concepts: The idea behind the Kuiper Belt was first proposed by Irish astronomer Kenneth Edgeworth in 1943, and the idea behind the Oort Cloud was brought up in 1932 by Estonian astronomer Ernst Öpik. For these reasons, these realms of the outer solar system are sometimes referred to as the Edgeworth-Kuiper Belt and the Öpik-Oort Cloud.

9 As told by Luu to Davies, *Beyond Pluto*, 50.

10 Such provisional names reflect the year of discovery (1992), the half-month during which the discovery was announced (Q denotes the first half of September), and the object's place on the chronological list for that half-month (A through Z, then A_1 through Z_1, A_2 through Z_2, and so on).

11 Davies, *Beyond Pluto*, 71.

12 Clyde W. Tombaugh, "Pluto: The Final Word," *Sky & Telescope* 88, no. 6 (December 1994): 8–9.

13 William Dicke, "Clyde W. Tombaugh, 90, Discoverer of Pluto," *New York Times*, January 20, 1997, C15.

14 David H. Freedman, "When Is a Planet Not a Planet?" *Atlantic Monthly* 281, no. 2 (February 1998): 22–33.

15 "Send Those Scientists to Pluto," editorial, *Peoria Journal Star*, January 31, 1999, p. A6.

16 Kurt Loft, "Planet 9 from Outer Space," *Tampa Tribune*, February 22, 1999, 4.

17 DPS Mailing 99-03 (January 28, 1999), http://dps.aas.org/mail_archive/dpsm.99-03, accessed May 28, 2009.

18 Stephen P. Maran, "Pluto: What It Is," *Washington Post*, March 10, 1999, H5.

19 John Fleck, "Tiny, Misfit Pluto May Lose Planet Status," *Albuquerque Journal*, January 19, 1999, A1.

20 Associated Press, "Town Residents Furious about the Possible Demotion of Pluto," January 27, 1999.

21 Frank Roylance, "Reclassify Planet Pluto? Defenders Go into Orbit; 'War of the Worlds' Erupts over Labeling Orb as a Lesser Light," *Baltimore Sun*, January 26, 1999, A1.

22 Johannes Andersen, "The Status of Pluto: A Clarification," IAU Press Release 01/99 (February 3, 1999), http://nssdc.gsfc.nasa.gov/planetary/text/pluto_iau_pr_19990203.txt, accessed May 28, 2009.

23 Jessica Dayton, "'Pluto Pride' Evident at Cosmic Celebration," *Peoria Journal Star*, February 23, 1999, B1.

24 Neil deGrasse Tyson, *The Pluto Files: The Rise and Fall of America's Favorite Planet* (New York: W.W. Norton & Co., 2009), 110.

25 Michelle Roberts, "75 Years After Discovery, Pluto Still Puzzles," Associated Press, February 14, 2005, http://www.msnbc.msn.com/id/6968617/, accessed May 28, 2009.

7. Not Yet Explored

1 S. Alan Stern, "The New Horizons Pluto Kuiper Belt Mission: An Overview with Historical Context," *Space Science Reviews* 140, nos. 1–4 (October 2008): 5–6.

2 Alan Stern and Jacqueline Mitton, *Pluto and Charon: Ice Worlds on the Ragged Edge of the Solar System*, 2nd ed. (Weinheim, Germany: Wiley-VCH, 2005), 182–183. In addition to Stern, the founding members included Fran Bagenal, Rick Binzel, Marc Buie, Bob Marcialis, Bill McKinnon, Ralph McNutt, Bob Millis, Ed Tedesco, Larry Trafton, Larry Wasserman, and Roger Yelle.

3 Ibid., 213.

4 Tom McNichol, "Beyond Cool," *Wired* 9.04 (April 2001): 116–128.

5 Stern and Mitton, *Pluto and Charon*, 215.

6 Brian Berger, "Congress Passes NASA Budget; Saves Pluto Mission," *Space News*, November 12, 2001, http://www.space.com/spacenews/Budget_1112.html, accessed May 29, 2009.

7 *New Frontiers in the Solar System: An Integrated Exploration Strategy*, National Research Council Solar System Exploration Survey (Washington, DC: National Academies Press, 2003), 4.

8 Bruce Moomaw, "The Bizarre 'Pluto War' Is Almost Over at Last, and Pluto Is Winning," *Space Daily*, October 9, 2002, http://www.spacedaily.com/news/outerplanets-02m.html, accessed May 29, 2009.

9 New Horizons' radioisotope thermoelectric generator is not a nuclear reactor, but rather a device that converts the heat created by 11 kilograms (24 pounds) of plutonium dioxide into electricity.

10 Robert Pearlman, "To Pluto, with Postage: Nine Mementos Fly with NASA's First Mission to the Last Planet," CollectSpace, October 28,

2008, http://www.collectspace.com/news/news-102808a.html, accessed May 29, 2009.

11 "Pluto-Bound, Student-Built Dust Detector Renamed 'Venetia,' Honoring Girl Who Named Ninth Planet," press release from New Horizons, Johns Hopkins University Applied Physics Laboratory, June 29, 2006, http://pluto.jhuapl.edu/news_center/news/062906.html, accessed May 29, 2009.

12 William Grimes, "Venetia Phair Dies at 90; as a Girl, She Named Pluto," *New York Times*, May 11, 2009, A21.

13 H. A. Weaver et al., "The Discovery of Two New Satellites of Pluto," *Nature* 439 (February 23, 2006): 943–945. Hydra was the serpent of the underworld in Greek mythology (killed by Heracles), and Nix was the goddess of the night (the Egyptian spelling of the name was used to avoid confusion with the asteroid Nyx, which was named using the traditional Greek spelling).

14 Tariq Malik, "Reaching for Pluto: NASA Launches Probe to Solar System's Edge," Space.com, January 19, 2006, http://www.space.com/mission launches/060119_pluto_nh_launch.html, accessed May 29, 2009.

8. Betting on the Tenth Planet

1 Kenneth Chang and Dennis Overbye, "Planet or Not, Pluto Now Has Far-out Rival," *New York Times*, July 30, 2005, A1.

2 Robert Roy Britt, "Astronomers Believe More Planets Lie Beyond Kuiper Belt," *Space News*, December 8, 2004, http://www.space.com/spacenews/archive04/kuiperarch_120604.html, accessed May 29, 2009.

3 Cal Fussman, "The Man Who Finds Planets," *Discover*, May 2006, http://discovermagazine.com/2006/may/cover, accessed May 29, 2009.

4 Ibid.

5 Hubble Space Telescope measurements put its diameter at about 800 miles, compared with Pluto's diameter of 1,430 miles. Quaoar's mass is less than a tenth of Pluto's mass.

6 Stephen Cauchi, "Quaoar, the Newest Planet . . . Or Is It?" *The Age* (Sydney, Australia), October 8, 2002, 1.

7 S. J. Desch, "Mass Distribution and Planet Formation in the Solar Nebula," *Astrophysical Journal* 671 (December 10, 2007): 878–893.

8 Alan Stern was among the first to use the term "ice dwarfs," as quoted in Ron Cowen, "Plutos Galore," *Science News* 140, no. 12 (September 21, 1991): 184–186.

9 Govert Schilling, *The Hunt for Planet X: New Worlds and the Fate of Pluto* (New York: Copernicus Books, 2009), 191.

10 Kathy A. Svitil, "Farthest, Coldest 'Planet' Spied Well Beyond Pluto," *Discover*, January 2005, http://discovermagazine.com/2005/jan/farthest-coldest-planet, accessed May 29, 2009.

11 David Jewitt has noted that many comets and at least one Kuiper Belt object reach a point farther from the sun at their greatest separation, but that the Flying Dutchman (a.k.a. Sedna) is notable because its closest distance from the sun is still so far away. Eris, the icy world that caused so much trouble for the IAU, is currently farther away than Sedna, but that will change over time.

12 Michael Brown, Chadwick Trujillo, and David Rabinowitz, "Discovery of a Candidate Inner Oort Cloud Planetoid," *Astrophysical Journal* 617 (December 10, 2004): 617–645, http://www.gps.caltech.edu/~mbrown/papers/ps/sedna.pdf, accessed May 29, 2009.

13 Alessandro Morbidelli and Harold F. Levison, "Scenarios for the Origin of the Orbits of the Trans-Neptunian Objects 2000 CR105 and 2003 VB12 (Sedna)," *Astronomical Journal* 128 (2004): 2564–2576.

14 Alec Wilkinson, "The Tenth Planet," *New Yorker* 82, no. 22 (July 24, 2006): 50.

15 Fussman, "The Man Who Finds Planets." Wilkinson's *New Yorker* profile quotes Brown as saying something slightly different: "The first thing I do is pick up the phone and call my wife and say, 'I just won the bet.'"

16 At one point, Brown had suggested to the IAU that Xena be given the name Lilah, which would formally refer to Hindu concept of the universe as a puppet theater (but informally refer to his daughter). Brown's wife reportedly persuaded him to find another name, telling him, "What if we have a second child? You'd have to go find another planet." (Kenneth Chang, "Dwarf Planet, Cause of Strife, Gains 'the Perfect Name,'" *New York Times*, September 15, 2006, A20.)

17 Robert Roy Britt, "Object Bigger than Pluto Discovered, Called 10th Planet," Space.com, July 29, 2005, http://www.space.com/scienceastronomy/050729_new_planet.html, accessed May 29, 2009.

9. The Battle of Prague

1 Laurel Kornfeld, "Pluto, the Planet That Was," Space.co.uk (2008), http://www.space.co.uk/Features/Articles/Plutotheplanetthatwas/tabid/634/Default.aspx, accessed May 29, 2009.

2 Xena, now known as Eris, is brighter in absolute terms—that is, if Pluto and Eris were the same distance from Earth, Eris would appear brighter. Technically speaking, Eris's absolute magnitude is −1.20, while Pluto's absolute magnitude of −0.81 would make it dimmer. Mercury would be dimmer still, with an absolute magnitude of −0.40.

3 Steven Soter, "What Is a Planet?" *Astronomical Journal* 132 (December 2006): 2513–2519; Table 2: "Primary Reservoirs of Colliding Objects."

4 Alan Boss, *The Crowded Universe: The Search for Living Planets* (New York: Basic Books, 2009), 121.

5 Owen Gingerich, "The Path to Defining Planets," *Dissertatio Cum Nuncio Siderio III*, August 16, 2006, 4.

6 If you apply the criterion in the initial draft of the IAU planet definition, this would occur when the Earth-moon system's center of mass, or barycenter, was somewhere between the surface of the Earth and the moon, rather than beneath Earth's surface as it is now. This won't happen for billions of years. The definition applied by the IAU rules out having any double planets.

7 Unless otherwise noted, quotations from the IAU sessions are taken from the IAU's video archive of those sessions, available online via http://www.astronomy2006.com/media-stream-archive.php, accessed May 29, 2009.

8 Dennis Overbye, "Astronomers in a Quandary over Pluto's Status," *New York Times*, August 23, 2006, A20.

9 Owen Gingerich, "Planetary Perils in Prague," *Daedalus* 136, no. 1 (Winter 2007): 140.

10 Jim Erickson, "Pluto Booted Off Roster of Planets," *Rocky Mountain News* (Denver), August 25, 2006, A4.

10. The Lighter Side of Pluto

1 Donna Smith, "Pluto Becomes One Less Planet to Memorize," Reuters, August 24, 2006, http://www.redorbit.com/news/science/631205/pluto_becomes_one_less_planet_to_memorize/index.html, accessed May 29, 2009.

2 Editorial cartoon by Robert Ariail for *The State* (Columbia, SC), August 25, 2006, http://www.cagle.com/news/Pluto/3.asp, accessed May 29, 2009.

3 Transcript of jokes from late-night comics on August 28, 2006, http://my.cnd.org/modules/newbb/viewtopic.php?topic_id=32931&forum=1&start=1100&viewmode=flat&order=0, accessed May 29, 2009.

4 "Poor Pluto," http://www.youtube.com/watch?v=0w0hpyNenjM, accessed May 29, 2009.

5 Quotations from "Pluto: Good or Bad for the Jews?" *The Jewish Angle*, September 2006, http://jangle04.home.mindspring.com/1033.html, accessed May 29, 2009.

6 "Dwarf Planet, Dwarf President Button," http://www.zazzle.com/dwarf_planet_dwarf_president_button-145621125082185557, accessed May 29, 2009.

7 Tom Teepen, "Astronomers Goofed on Pluto," *Albany (NY) Times Union*, August 30, 2006, A15.

8 Robert Roy Britt, "Pluto Demoted: No Longer a Planet in Highly Controversial Definition," August 24, 2006, http://www.space.com/scienceastronomy/060824_planet_definition.html, accessed May 29, 2009.

9 California Institute of Technology news release, "Xena Awarded 'Dwarf Planet' Status, IAU Rules; Solar System Now Has Eight Planets," August 24, 2006, http://mr.caltech.edu/press_releases/12886, accessed May 29, 2009.

10 Reuters, "World Responds to Pluto without Planet Tag," August 24, 2006, and Jim Erickson, "Pluto Booted Off Roster of Planets," *Rocky Mountain News*, August 25, 2006, A4.

11 William J. Kole, "Dinky Pluto Loses Its Status as Planet," Boston Globe, Associated Press dispatch, August 24, 2006, http://www.boston.com/news/world/europe/articles/2006/08/24/astronomers_say_pluto_is_not_a_planet/, accessed May 29, 2009.

12 Sarah Zielinski, "In Brief: New Mexico Declares Pluto a Planet," *Eos: Transactions of the American Geophysical Union* 88, no. 11.

13 Legislative action detailed online at http://legistar.cityofmadison.com/DetailReport/?key=5246, accessed May 29, 2009.

14 Assembly House Resolution 36, introduced August 24, 2006, by members Richman and Canciamilla, http://info.sen.ca.gov/pub/05-06/bill/asm/ab_0001-0050/hr_36_bill_20060824_introduced.html, accessed May 29, 2009.

15 Cho Ji-hyun, "Cloned Cow, Pluto to Be Removed from School Texts," *Korea Herald*, April 9, 2007.

16 "The Outer Planets," *Earth Science Study Guide*, Section 27-2, http://timhayes.com/earthsci/StudyGuides/Chapter27StudyGuide.pdf, accessed May 29, 2009.

17 Carol Scott, "School Day: Pluto Erased from Lessons," *Newport News DailyPress*, October 30, 2006, C1.

18 Jeanna Bryner, "Pluto's Identity Crisis Hits Classrooms and Bookstores," Space.com, June 19, 2008, http://www.space.com/scienceastronomy/080619-pluto-confusion.html, accessed May 29, 2009.

19 David A. Aguilar, *11 Planets: A New View of the Solar System* (Washington, DC: National Geographic Society, 2008). The eleven-planet mnemonic was suggested by Maryn Smith, the winner of a National Geographic contest.

20 Bryner, "Pluto's Identity Crisis."

21 "Air and Space Museum Demotes Pluto: Update," *Demote Pluto*, October 29, 2006, http://demotepluto.blogspot.com/2006/10/air-and-space-museum-demotes-pluto.html, accessed May 29, 2009.

22 Jason Kottke, "Bringing Pluto Back to the Solar System," Kottke.org, September 10, 2008, http://kottke.org/08/09/bringing-pluto-back-to-the-solar-system, accessed May 29, 2009.

23 Ashley Yeager, "From Planet to Plutoid," *Science News* 174, no. 1 (July 5, 2008): 7, expanded version online at http://www.sciencenews.org/view/generic/id/33158/title/From_planet_to_plutoid, accessed May 29, 2009. That the IAU's Executive Committee was unwilling to create a new "ceroid" category may have been an indication that the leadership was getting sick of the whole business of planet/nonplanet definition.

24 David Jewitt, "Plutoids, Rutoids, Schmutoids," June 11, 2008, http://www.ifa.hawaii.edu/faculty/jewitt/kb/plutoids.html, accessed May 29, 2009.

25 Rachel Courtland, "Pluto-like Objects to Be Called 'Plutoids,'" *New Scientist*, June 11, 2008, http://www.newscientist.com/article/dn14118-plutolike-objects-to-be-called-plutoids.html, accessed May 29, 2009.

11. The Great Planet Debate

1 Some have claimed that the IAU voted nearly unanimously to remove Pluto from the planet category. However, ninety-one votes were recorded in favor of Resolution 5B, which would have been interpreted as retaining

Pluto and other dwarf planets in the "big tent" of planethood. The nay votes were left untallied, but the implication is that about a quarter of the IAU voters wanted dwarf planets to be considered planets, based on the tallies for the IAU's only fully recorded votes, on Resolutions 6A and 6B.

2 Quoted in Brittany Graham (producer) and Colin Campbell (writer/ director), *The Universe: The Outer Planets*, originally aired August 14, 2007, on the History Channel.

3 In fact, the leading scientists behind the planet definition cooked up in Prague included specialists on comets, asteroids, the Kuiper Belt, and planetary migration.

4 Mark V. Sykes, "The Planet Debate Continues," *Science* 319, no. 5871 (March 28, 2008): 1765.

5 S. Alan Stern and Harold F. Levison, "Regarding the Criteria for Planethood and Proposed Planetary Classification Schemes," in *Highlights of Astronomy*, vol. 12, as presented at the XXIVth General Assembly of the IAU—2000 [Manchester, UK, August 7–18, 2000], edited by H. Rickman (San Francisco: Astronomical Society of the Pacific, 2002), 205–213.

6 Steven Soter, "What Is a Planet?" *Astronomical Journal*, 132 (December 2006): 2513–2519. Soter also explains his view in an article with the same title in the January 2007 issue of *Scientific American*.

7 Gonzalo Tancredi and Sofia Favre, "Which Are the Dwarfs in the Solar System?" *Icarus* 195, no. 2 (June 2008): 851–862. The candidate list included Ixion, Huya, Quaoar, Sedna, Orcus, Varuna, Haumea, and the numbered minor planets 15874, 55565, and 55636. Less clear candidates were 26375, 42301, 47171, 90568, 2001 QF_{298}, and 2003 AZ_{84}. The researchers have established a Web site, http://www.astronomia.edu .uy/dwarfplanet/, to provide updates on the candidate list.

8 Easter Island is becoming more commonly known by its Polynesian name, Rapa Nui.

9 Mike Brown, "What's In a Name? [Part 2]," posted online to *Mike Brown's Planets*, July 13, 2008, http://www.mikebrownsplanets .com/2008/07/whats-in-name-part-2.html, accessed May 30, 2009.

10 The moon formerly known as Rudolph was named Hi'iaka, after one of Haumea's daughters in Hawaiian mythology. Months after Santa and Rudolph were discovered, Brown's team found that Santa had a

second, smaller moon, which was nicknamed Blitzen, in keeping with the Christmas theme. In 2008, that moon was officially named Namaka, after another one of Haumea's children.

11 Rachel Courtland, "Controversial Dwarf Planet Finally Named 'Haumea,'" *New Scientist*, September 19, 2008, http://www.newscientist.com/article/ dn14759, accessed May 30, 2009.

12. Day of the Dwarfs

1 The smallest known galaxy is Willman 1, with the mass of 500,000 suns. That's so small that some astronomers suspect it's a globular star cluster rather than a galaxy. The largest known spiral galaxy is ISOHDF 27, with the mass of more than a trillion suns. The mass of the Abell 2029 galaxy cluster has been estimated at as much as 180 trillion suns, although there is some debate over whether this counts as a single galaxy known as cD or as a multiple cluster. The spread for stars is smaller: The theoretical limit for the most massive star is 150 solar masses, while the smallest known star is 0.02 solar masses.

2 As noted previously, Stern referred to "ice dwarfs" in Ron Cowen, "Plutos Galore: Ice Dwarfs May Dominate the Solar System's Planetary Population," *Science News* 140, no. 12 (September 21, 1991): 184.

3 Mike Brown, comment in "Haumea," *Mike Brown's Planets*, September 17, 2008, http://www.mikebrownsplanets.com/2008/09/haumea.html, accessed May 31, 2009.

4 D. T. Britt, G. J. Consolmagno, and W. J. Merline, "Small-Body Density and Porosity: New Data, New Insights," *Publications of the Lunar and Planetary Science Conference*, 2006 (Abst. 2214).

5 E. Lellouch, B. Sicardy, C. de Bergh, H.-U. Käulf, S. Kassi, and A. Campargue, "Pluto's Lower Atmosphere Structure and Methane Abundance from High-Resolution Spectroscopy and Stellar Occultations," *Astronomy & Astrophysics* 495, no. 3 (March 2009), L17-L21.

6 Mike Brown, "Haumea."

7 "Fourth Dwarf Planet Named Makemake," news release from the International Astronomical Union, July 19, 2008, http://www.iau.org/ public_press/news/release/iau0806/, accessed May 31, 2009.

8 Ray Villard, "An Escapee from the Outermost Solar System?" *Discovery Space: Cosmic Ray*, July 23, 2008, http://blogs.discovery.com/cosmic_ray/ 2008/07/an-escapee-from.html, accessed May 31, 2009.

9 Quoted in "Largest Asteroid May Be 'Mini-Planet' with Water Ice," news release from Space Telescope Science Institute, September 7, 2005, http://hubblesite.org/newscenter/archive/releases/2005/27/text/, accessed May 31, 2009.

10 Bruce Moomaw, "Ceres as an Abode of Life," *Space Daily*, July 2, 2007, http://www.spacedaily.com/reports/Ceres_As_An_Abode_Of_Life_999.html, accessed May 31, 2009.

11 The IAU calls for using the dwarf-planet naming process if an object's absolute magnitude is +1 or brighter. Charon's brightness falls at that boundary. If later evidence indicates that a dimmer celestial body is round, it would then be given dwarf-planet status, but the name would not be changed.

12 Leslie Mullen, "Delayed Gratification Zones," *Astrobiology Magazine*, April 7, 2004, http://www.astrobio.net/news/article912.html, accessed May 31, 2009.

13. Planet X Redux

1 A scholar in Semitic languages, Michael S. Heiser, points out that the word "Nibiru" is used in several different contexts in Sumerian cuneiform writings. The word is sometimes used to refer to the planet Jupiter—or, in one quotation, Mercury—but not to a world beyond the five naked-eye planets familiar to the ancients. ("The Myth of a Sumerian 12th Planet," http://www.michaelsheiser.com/nibiru.pdf, accessed May 31, 2009).

2 The science behind the controversy is explained in "No Tenth Planet Yet from IRAS," http://web.ipac.caltech.edu/staff/tchester/iras/no_tenth_planet_yet.html, accessed May 31, 2009.

3 David Morrison, "The Myth of Nibiru and the End of the World in 2012," *Skeptical Inquirer* 32, no. 5 (September–October 2008): 50–55.

4 With an orbit that ranges from 76 AU to 976 AU—well beyond the outer edge of the Kuiper Belt at about 50 AU—Sedna is considered the solar system's first known distant detached object. Other detached objects have been found since Sedna's discovery in 2003.

5 P. S. Lykawka, and T. Mukai, "An Outer Planet Beyond Pluto and the Origin of the Trans-Neptunian Belt Architecture," *Astronomical Journal* 135 (April 2008): 1161–1200. Lykawka and Mukai propose a planet with a semimajor axis of 100–175 AU, perihelion of at least 80 AU, mass of

30 percent to 70 percent of Earth's mass, with orbital inclination of 20 to 40 degrees. Orbital period would be 1,000 to 2,300 years.

6 Rodney S. Gomes, John J. Matese, and Jack J. Lissauer, "A Distant Planetary-Mass Solar Companion May Have Produced Distant Detached Objects," *Icarus* 184, no. 2 (October 2006): 589–601. The authors say distant detached objects could be produced by an Earth-mass object at 1,000 AU, a Neptune-mass object at 2,000 AU, or a Jupiter-mass object at 5,000 AU or farther.

7 Robert Roy Britt, "Solar System Makeover: Wild New Theory for Building Planets," Space.com, July 9, 2002, http://www.space.com/scienceastronomy/solarsystem/planet_formation_020709-1.html, accessed May 31, 2009.

8 Geoff Ward, "Riddle of Planet X," *Western Daily Press* (Bristol, UK), December 23, 2005, 11.

9 "Giant Magellan Telescope Science Case," 12–13, http://www.gmto.org/sciencecase/GMT-ID-01404-GMT_Science_Case.pdf, accessed May 31, 2009.

10 Steven Chesley et al., "Cataloging and Characterizing the Small Bodies of the Solar System With LSST," poster presented at the 213th Meeting of the American Astronomical Society, Long Beach, CA, January 6, 2009 (460.17), http://www.lsst.org/files/docs/aas/2009/chesley_419_Jan08.pdf, accessed May 31, 2009.

11 David Jewitt, "Project Pan-STARRS and the Outer Solar System," *Earth, Moon and Planets* 92 (2003): 470–475.

12 This estimate assumes that the typical small Kuiper Belt object has an albedo, or reflectivity factor, of 10 percent, and that anything judged wider than 400 kilometers on that basis would count as a dwarf planet. However, the IAU uses a different criterion for naming dwarf planets, based on absolute brightness. As of this writing, the IAU lists five dwarf planets, while Brown lists forty-six (http://www.gps.caltech.edu/~mbrown/dwarfplanets/, accessed May 31, 2009).

13 Quoted in "Brian Marsden Says Eight (Planets) Are Enough," Planetary Society Radio, September 25, 2006, http://www.planetary.org/radio/show/00000203/, accessed May 31, 2009.

14 Robert Roy Britt, "Strange New Worlds Could Make Miniature Solar Systems," Space.com, June 5, 2006, http://www.space.com/scienceastronomy/060605_planemos.html, accessed May 31, 2009.

14. Alien Plutos

1 Geoffrey W. Marcy and R. Paul Butler, "Giant Planets Orbiting Faraway Stars," *Scientific American Presents: Magnificent Cosmos* 9, no. 1 (Spring 1998), http://astro.berkeley.edu/~gmarcy/sciam.html, accessed May 31, 2009.

2 "Scientists Announce Smallest Extra-Solar Planet Yet Discovered and Find Outer Limits of the Pulsar Planetary System," Penn State news release, February 7, 2005, http://www.science.psu.edu/alert/Wolszczan2-2005.htm, accessed May 31, 2009.

3 Alice C. Quillen, Alessandro Morbidelli, and Alex Moore, "Planetary Embryos and Planetesimals Residing in Thin Debris Discs," *Monthly Notices of the Royal Astronomical Society* 380, no. 4 (October 2007): 1642–1648.

4 William Atkins, "Hubble Space Telescope Observes When Planets Were Babies," *ITWire*, January 10, 2007, http://www.itwire.com/content/view/8487/1154/, accessed May 31, 2009.

5 Paul Kalas, James R. Graham, and Mark Clampin, "A Planetary System as the Origin of Structure in Fomalhaut's Dust Belt," *Nature* 435 (June 23, 2005): 1067–1070.

6 Paul Kalas et al., "Optical Images of an Exosolar Planet 25 Light-Years from Earth," *Science* 322, no. 5906 (November 28, 2008): 1345–1348.

7 Christian Marois et al., "Direct Imaging of Multiple Planets Orbiting the Star HR 8799," *Science* 322, no. 5906 (November 28, 2008): 1348–1352.

8 A. C. M. Correia et al., "The HARPS Search for Southern Extra-solar Planets. XVI. HD 45364, a Pair of Planets in a 3:2 Mean Motion Resonance," *Astronomy and Astrophysics* (accepted for publication November 6, 2008), retrieved online as preprint at http://arxiv.org/abs/0902.0597.

9 You can watch them cross at http://www.planetary.org/exoplanets/list.php?exo=HD+82943+b, accessed May 31, 2009.

10 Ken Croswell, "Extrasolar Neptune-Pluto Analogue Discovered," February 18, 2009, http://kencroswell.com/HD45364.html, accessed May 31, 2009.

11 Alan Boyle, "Looking for Other Earths? Here's a List," MSNBC.com, February 18, 2006, http://www.msnbc.msn.com/id/11427824/, accessed May 31, 2009.

12 D. Backman et al., "Epsilon Eridani's Planetary Debris Disk: Structure and Dynamics Based on Spitzer and Caltech Submillimeter Observatory Observations," *Astrophysical Journal* 690 (January 10, 2009): 1522–1538.

13 G. Fritz Benedict et al., "The Extrasolar Planet Epsilon Eridani b: Orbit and Mass," *Astronomical Journal* 132 (November 2006): 2206–2218.

14 Thayne Currie, Peter Plavchan, and Scott J. Kenyon, "A Spitzer Study of Debris Disks in the Young Nearby Cluster NGC 2232: Icy Planets Are Common around ~1.5-3 M Stars," *Astrophysical Journal* 688 (November 20, 2008): 597–615.

15 The suggestion to develop a planet classification system analogous to the well-established stellar classification system has come from many quarters, including astronomer Steven Soter and Ray Villard, news director for the Space Telescope Science Institute (http://blogs.discovery.com/cosmic_ray/2008/07/the-hitchhikers.html, accessed May 31, 2009).

16 More recently, additional types of stars and brown dwarfs have been added to the classification system, requiring tweaks in the mnemonic. One alternate phrase is "Oh, Be A Fine Girl, Kiss Me—Right Now, Sweetheart!" Another is "Oh, Be A Fine Girl, Kiss My Lips Tonight." But the following might be the most apt memory aid: "Only Bored Astronomers Find Gratification Knowing Mnemonics."

17 Robert Roy Britt, "Why Planets Will Never Be Defined," Space.com, November 21, 2006, http://www.space.com/scienceastronomy/061121_exoplanet_definition.html, accessed May 31, 2009.

18 G. Marcy and R. Paul Butler, "Definition of a Planet," November 6, 2000, http://exoplanets.org/defn.html, accessed May 31, 2009.

19 David A. Weintraub, *Is Pluto a Planet? A Historical Journey Through the Solar System* (Princeton, NJ: Princeton University Press, 2007), 230.

15. The Case for Pluto

1 Arthur H. Compton, "Are Planets Rare?" *Science* 72, no. 1861 (August 29, 1930): 219.

2 Michel Mayor quoted in "A Trio of Super-Earths," news release from European Southern Observatory, June 16, 2008, http://www.eso.org/public/outreach/press-rel/pr-2008/pr-19-08.html, accessed May 31, 2009.

3 The super-Earth planet CoRoT-Exo-7b orbits at a distance of 0.017 AU, compared with Mercury's mean distance of 0.39 AU. Fomalhaut b is

thought to be at a mean distance of 119 AU from its star, compared with Pluto's mean distance of 39.5 AU.

4 Kathy A. Svitil, "Beyond Pluto," *Discover* 25, no. 11 (November 2004): 45.

5 S/2008 S 1 was reported as a moon orbiting within Saturn's G ring in Circular No. 9023 from the IAU's Central Bureau for Astronomical Telegrams, dated March 3, 2009. Other ring-arc moons include Anthe and Methone. All three moons and their arcs are held in place gravitationally by the larger moon Mimas.

6 David Grinspoon, "This Is Not a Planet?" *Sky & Telescope* 117, no. 3 (March 2009): 20. Emphasis in the original.

7 Comments made during the Great Planet Debate, August 14, 2008, http://gpd.jhuapl.edu/debate/debateStream.php, accessed May 31, 2009.

8 Quoted in "Advocating More Planets," supplemental video for *400 Years of the Telescope: A Journey of Science, Technology and Thought* (Chico, CA: Interstellar Studios, 2009), http://www.400years.org/video/view_video.php?k=13, accessed May 31, 2009.

SELECTED
BIBLIOGRAPHY

Scientific papers and other references relevant to the text have been cited in the footnotes. Here is a smaller sampling of books, articles, and Internet resources that could serve as the basis for a reading list on the outer solar system and the history of the search for planets. Most of the books have been published (or updated) since the International Astronomical Union's Prague meeting in 2006. The articles have been selected to provide a sense of how the planet debate has evolved over the past quarter century.

Books for General Audiences

Boss, Alan. *The Crowded Universe: The Search for Living Planets.* New York: Basic Books, 2009.

Davies, John. *Beyond Pluto: Exploring the Outer Limits of the Solar System.* Cambridge, UK: Cambridge University Press, 2001.

Jones, Tom, and Ellen Stofan. *Planetology: Unlocking the Secrets of the Solar System.* Washington, DC: National Geographic, 2008.

Lemonick, Michael. *The Georgian Star: How William and Caroline Herschel Revolutionized Our Understanding of the Cosmos.* New York: W.W. Norton & Co., 2009.

Levy, David H. *Clyde Tombaugh: Discoverer of Planet Pluto*. Tucson: University of Arizona Press, 1991.

Littmann, Mark. *Planets Beyond: Discovering the Outer Solar System*. New York: John Wiley & Sons, 1990.

Marschall, Laurence A., and Stephen P. Maran. *Pluto Confidential: An Insider Account of the Ongoing Battles over the Status of Pluto*. Dallas, TX: Ben Bella Books, 2009.

McFadden, Lucy Ann Adams, Paul Robert Weissman, and Torrence V. Johnson, eds. *Encyclopedia of the Solar System* 2nd ed. San Diego: Academic Press, 2007.

Miller, Ron, and William K. Hartmann. *The Grand Tour: A Traveler's Guide to the Solar System*. 3rd ed. New York: Workman Publishing, 2005.

Minard, Anne, and Carolyn Shoemaker. *Pluto and Beyond: A Story of Discovery, Adversity, and Ongoing Exploration*. Flagstaff, AZ: Northland Publishing, 2007.

Schilling, Govert. *The Hunt for Planet X: New Worlds and the Fate of Pluto*. New York: Copernicus Books, 2009.

Sobel, Dava. *The Planets*. New York: Viking Penguin, 2005.

Standage, Tom. *The Neptune File: A Story of Astronomical Rivalry and the Pioneers of Planet Hunting*. New York: Walker & Co., 2000.

Stern, Alan, and Jacqueline Mitton. *Pluto and Charon: Ice Worlds on the Ragged Edge of the Solar System*. 2nd ed. Weinheim, Germany: Wiley-VCH, 2005.

Sutherland, Paul. *Where Did Pluto Go? A Beginner's Guide to Understanding the New Solar System*. Pleasantville, NY: Reader's Digest, 2009.

Tyson, Neil deGrasse. *The Pluto Files: The Rise and Fall of America's Favorite Planet*. New York: W.W. Norton & Co., 2009.

Weintraub, David A. *Is Pluto a Planet? A Historical Journey Through the Solar System*. Princeton, NJ: Princeton University Press, 2007.

Children's Books

Aguilar, David A. *11 Planets: A New View of the Solar System*. Washington, DC: National Geographic Society, 2008.

Carson, Mary Kay. *Extreme Planets! Q&A*. Smithsonian Q&A Series. New York: Collins, 2008.

Cole, Joanna. *The Magic School Bus: Lost in the Solar System*. Rev. ed. New York: Scholastic Press, 2006.

Croswell, Ken. *Ten Worlds: Everything That Orbits the Sun*. Rev. ed. Honesdale, PA: Boyds Mills Press, 2007.

Howard, Fran. *The Kuiper Belt*. Edina, MN: ABDO Publishing, 2008.

Landau, Elaine. *Beyond Pluto: The Final Frontier in Space*. New York: Children's Press, 2008.

———. *Pluto: From Planet to Dwarf*. New York: Children's Press, 2008.

Rusch, Elizabeth. *The Planet Hunter: The Story Behind What Happened to Pluto*. Flagstaff, AZ: Rising Moon, 2007.

Scott, Elaine. *When Is a Planet Not a Planet? The Story of Pluto*. New York: Clarion Books, 2007.

Wetterer, Margaret K. *Clyde Tombaugh and the Search for Planet X*. Minneapolis: Carolrhoda Books, 1996.

Winrich, Ralph, updated by Thomas K. Adamson. *Pluto: A Dwarf Planet*. Mankato, MN: Capstone Press, 2008.

Articles in Periodicals

Cowen, Ron. "Plutos Galore." *Science News* 140, no. 12 (September 21, 1991): 184–186.

Freedman, David H. "When Is a Planet Not a Planet?" *Atlantic Monthly* 281, no. 2 (February 1998): 22–33.

Fussmann, Cal. "The Planet Finder." *Discover* 27, no. 5 (May 2006): 38–45.

Gingerich, Owen. "Planet Politics: How I Tried—and Failed—to Save Pluto." *Boston Globe*, September 3, 2006, D2.

———. "Planetary Perils in Prague." *Daedalus* 136, no. 1 (Winter 2007): 137–140.

Grinspoon, David. "This Is Not a Planet?" *Sky & Telescope* 117, no. 3 (March 2009): 20.

Jewitt, David, and Jane X. Luu. "Pluto, Perception & Planetary Politics." *Daedalus* 136, no. 1 (Winter 2007): 132–136.

Malhotra, Renu. "Migrating Planets." *Scientific American* 281, no. 3 (September 1999): 56–63.

Marsden, B. G. "Planets and Satellites Galore." *Icarus* 44 (October 1980): 29–37.

Morrison, David. "The Myth of Nibiru and the End of the World in 2012." *Skeptical Inquirer* 32, no. 5 (September–October 2008): 50–55.

Robertson, William C. "Science 101: Why Is Pluto No Longer a Planet?" *Science and Children* 44, no. 3 (November 2006): 60–61.

Schilling, Govert. "Underworld Character Kicked Out of Planetary Family." *Science* 313, no. 5791 (September 1, 2006): 1214–1215.

Soter, Steven. "What Is a Planet?" *Astronomical Journal* 132 (December 2006): 2513–2519.

———. "What Is a Planet?" *Scientific American* 296, no. 1 (January 2007): 34–41.

Stern, S. Alan. "Debates Over Definition of Planet Continue and Inspire." *Science News* 174, no. 12 (December 6, 2008): 32.

———. "Journey to the Farthest Planet." *Scientific American* 286, no. 5 (May 2002): 56–63.

———. "The New Horizons Pluto Kuiper Belt Mission: An Overview with Historical Context." *Space Science Reviews* 140, nos. 1–4 (October 2008): 5–6.

Stern, S. Alan, and Harold F. Levison. "Regarding the Criteria for Planethood and Proposed Planetary Classification Schemes." In *Highlights of Astronomy*, vol. 12, as presented at the XXIVth General Assembly of the IAU—2000 (Manchester, UK, August 7–18, 2000). Edited by H. Rickman. San Francisco: Astronomical Society of the Pacific, 2002, 205–213.

Svitil, Kathy A. "Beyond Pluto." *Discover* 25, no. 11 (November 2004): 42–49.

Sykes, Mark V. "The Planet Debate Continues." *Science* 319, no. 5871 (March 28, 2008): 1765.

Tombaugh, Clyde W. "The Search for the Ninth Planet, Pluto." *Astronomical Society of the Pacific Leaflets*, no. 209 (July 1946): 73–80.

Wilkinson, Alec. "The Tenth Planet." *New Yorker* 82, no. 22 (July 24, 2006): 50–65.

Online Articles

Many of the articles listed in the previous section are available online as well, but the following articles have been published exclusively online. All Web pages were last accessed on March 22, 2009.

Britt, Robert Roy. "Pluto Demoted: No Longer a Planet in Highly Controversial Definition." Space.com, August 24, 2006, http://www .space.com/scienceastronomy/060824_planet_definition.html.

Brown, Michael. "Defining Planets." *Astrobiology Magazine*, February 22, 2007. Five-part transcript of lecture, http://www.astrobio.net/ news/article2249.html.

Fischer, Daniel. "Inside the Planet Definition Process." *The Space Review*, September 11, 2006, http://www.thespacereview.com/ article/703/1.

Fraknoi, Andrew, ed. "Teaching What a Planet Is: A Roundtable on the Educational Implications of the New Definition of a Planet." *Astronomy Education Review* 5, September 2006–May 2007, http:// aer.noao.edu/cgi-bin/article.pl?id=207.

Minor Planet Electronic Circular 1999-C03: Editorial Notice, February 4, 1999, http://www.cfa.harvard.edu/mpec/J99/J99C03.html.

Mullen, Leslie. "Delayed Gratification Zones." *Astrobiology Magazine*, April 7, 2004, http://www.astrobio.net/news/article912.

Naeye, Robert. "The Case for Pluto." *Sky & Telescope Homepage Blog*, September 6, 2006, http://www.skyandtelescope.com/community/ skyblog/home/3850947.html.

O'Neill, Ian. "2012: No Planet X." *Universe Today*, May 25, 2008, http:// www.universetoday.com/2008/05/25/2012-no-planet-x/.

Pearlman, Robert Z. "To Pluto, with Postage: Nine Mementos Fly with NASA's First Mission to the Last Planet." *CollectSpace*, October 28, 2008, http://www.collectspace.com/news/news-102808a.html.

Pullen, Lee. "Could Life on Earth Have Come from Ceres?" *Astrobiology Magazine*, March 5, 2009, http://www.astrobio.net/news/article3058.

Than, Ker. "Pluto-Sized Planet Embryos Detected." Space.com, October 1, 2007, http://www.space.com/scienceastronomy/071001_mm_ planet_embryos.html.

Other Internet Resources

The Case for Pluto on the Web: http://www.thecaseforpluto.com/
Dwarf Planet & Plutoid Headquarters (Uruguay): http://www.astro nomia.edu.uy/dwarfplanet/

Dwarf Planets (Caltech): http://web.gps.caltech.edu/~mbrown/
    dwarfplanets.html/

The Great Planet Debate: http://gpd.jhuapl.edu/

IAU video archive for the XXVIth General Assembly: http://www
    .astronomy2006.com/media-stream-archive.php/

Mike Brown's Planets: http://www.mikebrownsplanets.com/

New Horizons: http://pluto.jhuapl.edu/

Pluto Today: http://www.plutotoday.com/

Plutonian News Network: http://www.plutoisaplanet.com/

CREDITS

Text page ii: Darren Phillips, New Mexico State University; pages 28, 37, 41: Lowell Observatory; page 55, U.S. Naval Observatory; pages 65, 131, 132: International Astronomical Union; page 90: courtesy of Wilbur Sitze; page 99: photo by Mike Brown's mom; page 151: courtesy of Michael Cameron; page 153: David Britt-Friedman, msnbc.com

Insert page 1: (top) NASA, ESA, H. Weaver (JHU/APL), A. Stern (SwRI), and the HST Pluto Companion Search Team; (bottom) Eliot Young et al., Southwest Research Institute, NASA; page 2: Dan Durda (Southwest Research Institute), Johns Hopkins University Applied Physics Laboratory; page 3: (top) NASA, ESA, and M. Brown, California Institute of Technology; (bottom) International Astronomical Union; page 4: Don Foley/Newsinfographics; page 5: (top) NASA, ESA, P. Kalas, J. Graham, E. Chiang, E. Kite (UC–Berkeley), M. Clampin (NASA GSFC), M. Fitzgerald (LLNL), and K. Stapelfeldt and J. Krist (NASA JPL); (bottom) for Ceres: NASA, ESA, L. McFadden and J. Y. Li (UMCP), M. Mutchler and Z. Levay (STScl), P. Thomas (Cornell), J. Parker and E. F. Young (SwRI), C. T. Russell and B. Schmidt (UCLA); for Vesta: NASA, ESA, J. Parker (SwRI), P. Thomas (Cornell), L. McFadden and J. Y. Li (UMCP), M. Mutchler and Z. Levay (STScl); page 6: Martin Kornmesser, International Astronomical Union; page 7: based on illustrations by NASA, ESA, and A. Field (STScl); page 8: NASA

INDEX

Note: Page numbers in *italics* refer to photos.